10667662

BP BLOWOUT

BP BLOWOUT

INSIDE THE GULF OIL DISASTER

DANIEL JACOBS

BROOKINGS INSTITUTION PRESS
Washington, D.C.

Copyright © 2016
Daniel Jacobs

All rights reserved. No part of this publication may be
reproduced or transmitted in any form or by any means without
permission in writing from the Brookings Institution Press.

The Brookings Institution is a private nonprofit organization
devoted to research, education, and publication on important
issues of domestic and foreign policy. Its principal purpose
is to bring the highest quality independent research and
analysis to bear on current and emerging policy problems.
Interpretations or conclusions in Brookings publications
should be understood to be solely those of the authors.

Library of Congress Cataloging-in-Publication data are available.
ISBN 978-0-8157-2908-2 (cloth : alk. paper)
ISBN 978-0-8157-2909-9 (ebook)

9 8 7 6 5 4 3 2 1

Typeset in Sabon and Gotham Condensed

Composition by Cynthia Stock

CONTENTS

PREFACE

In 2010 a blowout occurred at one of BP's deepwater wells in the Gulf of Mexico, killing 11 people and causing more than 3 million barrels of oil to hemorrhage into the ocean waters over 87 days. The oil that surfaced and washed ashore across the five Gulf Coast states affected the lives of millions of people. The oil also killed tens of thousands of wildlife and caused severe damage to natural resources. By mid-2016 BP had set aside nearly $62 billion to cover the costs of the end results—no small amount even for one of the richest companies in the world.

This is the inside story of the disaster and its aftermath. It is a story that has emerged more fully with time. Corporate homicide. The largest offshore oil discharge and the worst environmental disaster in American history. The most expensive manmade corporate disaster on record. All at the hands of one of the largest multinational corporations in the world. And all avoidable.

Although primary fault lies with BP, the federal government failed in its responsibility to oversee offshore drilling. To make matters worse, neither BP nor the federal government was prepared to cope with such a massive blowout. Both were ill-equipped to stop the flow of the oil from the well or to contain it once it reached the surface—efforts on which BP reported spending more than $14 billion. The company and the government struggled with damage control on two fronts: the actual devastation and the perceived harm to their images.

The legal fallout was immense. Hundreds of thousands of private claims poured in, and the courts were flooded with lawsuits. The claims process became contentious, leaving both private parties and BP feeling cheated at times. Individuals and businesses who filed claims against BP for economic and property losses ultimately were awarded a total of more than $13.5 billion.

The federal government brought criminal charges against BP and four of its employees. The company pled guilty to manslaughter and other charges to resolve the criminal case, agreeing to pay a record $4 billion in fines and penalties. Two BP employees were acquitted, and two pled guilty to misdemeanors. No one will go to prison for the accident. In stark contrast, the federal government prosecuted hundreds of individuals for filing false claims against BP. Seventy-five were incarcerated.

The federal government also brought a civil suit against BP. The case was tried in New Orleans in three separate phases between 2013 and 2015. A few months after the trial ended, the civil suit settled for over $20 billion, much of which will go to the Gulf.

BP has paid a dear price financially for the disaster—a total of nearly $62 billion—more than any company has ever paid for a disaster of its making. Society also has incurred substantial costs, although these are more difficult to quantify precisely.

What lessons can be learned? From a broad business perspective, the disaster teaches us about the need for principled leadership, state-of-the-art risk management, and progressive sustainability practices. A company like BP could have afforded to have been the best in its class—to have been a model in all three areas. It was not.

From a public policy perspective, the disaster teaches us that we need to take a fresh look at how much offshore oil drilling we allow in U.S. waters, especially deep water. We also need to develop a better means of deterring corporate conduct that gambles with people's lives, livelihoods, health, and shared natural resources. Oil companies that abuse the privilege of drilling in U.S waters should risk losing that privilege, partially or entirely.

The BP Gulf oil disaster provides an important case study with equally valuable lessons in business, sustainability, and government.

ACKNOWLEDGMENTS

I began and finished this book in California, my new home. Thanks to UCLA's Institute of the Environment and Sustainability, which hosted me twice as a visiting scholar, I was able to research and write in the company of distinguished scholars and amicable colleagues. I am especially grateful to Magali Delmas, director of the Institute's Center for Corporate Environmental Performance, and to Mark Gold, UCLA's associate vice chancellor for environment and sustainability. Heather Elms (my heroine) and Mark Clark, my former East Coast colleagues, provided invaluable mentorship over the years.

I am indebted to friends who generously took time out from their own busy schedules to support my effort in a myriad of ways. Ellen Spitalnik, Nicolas Kublicki, Mark Starik, and Murray Dry read and commented on early drafts. Jim Bayuk was not only a great sounding board, but also an unwavering source of reason and encouragement at every juncture. Susan Spinrad Esterly was both a superb listener and purveyor of sage wisdom.

My students have always been an inspiration to me. I was fortunate enough to have had the help of a number of excellent short-term student research assistants who managed to fit my work in with their studies. Leryn Gorlitsky, my long-term professional research assistant and now good friend, deserves special credit.

Some of the most interesting parts of the story would have been lost without my sources. Bill Reilly, my hero since the Rio Earth

Summit, did not disappoint, and I became an instant admirer of Bob Graham. Jane Lubchenco, who invited me to her home on a Saturday morning, and Bob Bea, who stood outside a UC–Berkeley parking lot to welcome me, overwhelmed me with their graciousness and their candor. So did others who prefer not to be named.

Last, I could never have gotten as far with this challenging endeavor (or in life) without the support of my loving family—immediate and extended.

Thank you all.

BP BLOWOUT

1

MAYDAY IN THE GULF

At 9:53 P.M. on April 20, 2010, Andrea Fleytas sent a "Mayday" signal from the *Deepwater Horizon*, a mobile oil rig sitting some 50 miles off the coast of Louisiana in the Gulf of Mexico. The rig was connected to a BP oil well a mile down on the ocean's floor. The well had suffered a blowout. When a well blows out, it can mean total loss of control, just like when a tire blows out on a car traveling at high speed. Fluids and natural gas shot up from the well, causing an explosion on board the rig, which became engulfed in flames. Disaster had struck.

Fleytas, a 23-year-old junior bridge officer, was a 2008 graduate of the California Maritime Academy. This was her first job on a vessel. She later reported that when she told the rig's captain about the distress call, he turned to her and cursed, asking: "Did I give you authority to do that?"[1]

Given the dire circumstances on the *Deepwater Horizon,* Fleytas did not jump the gun in sending for help. The Mayday signal was relayed to the U.S. Coast Guard. It sent two vessels, a rescue plane, and four helicopters. By the time they arrived at the scene, their rescue effort was futile. Eleven people were dead, most likely as a result of the explosion itself. The remaining 115 crew members, some seriously injured, survived. Two of the survivors were high-level BP executives. In an ironic coincidence, they were on board the *Deepwater Horizon* to give its crew an award for safety.

Two days later, Earth Day, the *Deepwater Horizon* sank. In the process, the pipe connecting it to the well snapped. Oil from the well started to spout uncontrollably into the Gulf's waters. For days, then weeks, then months.

Nearly two months later, President Barack Obama gave his first Oval Office address ever, calling the incident the nation's "worst environmental disaster."[2] By then, June 15, the discharge of oil already far exceeded the 11 million gallon total that flowed from the *Exxon Valdez* tanker after it hit a reef in Alaskan waters in 1989. By the time the BP well was finally capped on July 15, 2010, 134 million gallons (3.19 million barrels) of oil had spewed into the ocean, nearly surpassing the world record of 147 million gallons from the Ixtoc well in Mexican waters in the Gulf in 1979.

What happened was also the nation's worst sustainability disaster. Sustainability typically includes three pillars: social, environmental, and economic. The consequences of the accident were devastating from all three perspectives. In addition to taking the lives of 11 people, the disaster affected the lives, livelihoods, and health of millions more. The oil also caused severe environmental damage. Depending on how the total cost of the disaster is calculated, it could approach a far higher number than the $62 billion BP is expected to pay out.

This is the story of that disaster. The word "disaster" is used here instead of "spill" because spill is much too mild a term for what happened. Bob Bea, a disaster expert, a distinguished professor emeritus at the University of California Berkeley School of Engineering (who once worked on oil rigs and also as a BP consultant), has used much more graphic language: "I call it a massive cluster f___."[3]

Whatever it is called, with the passage of time, we now know more than ever before about the disaster—what caused it and its social, environmental, and economic ramifications. We now have the benefit of extensive evidence from a lengthy federal court trial, as well as scientific and medical studies and financial data. And there is new information from people who want the story told as completely as possible, including those who worked for BP and the federal government. It is an important story for anyone concerned with sustainability, whether in their day-to-day life, the classroom, or the boardroom.

WHAT WENT WRONG AND WHY?

What caused the worst sustainability disaster in American history? Human error. Lots of it. Enough to constitute "gross negligence," according to the federal judge who presided over the civil trial and wrote a lengthy decision explaining BP's mistakes.[4] One of the world's largest oil companies and multinational corporations badly mismanaged an offshore oil venture.

Deepwater drilling can be highly lucrative but also very dangerous. As any scuba diver knows, pressure increases with depth. Deep water therefore brings higher risk. And this wasn't just deep water, it was ultra-deep water. The wellhead was one mile below the surface of the ocean, and the bottom of the well was another two miles down. Most wells in the Gulf are located in much more shallow water (1,000 feet or less).

Despite the inherent danger of this type of offshore drilling, BP repeatedly made decisions that made the project substantially riskier:

—BP cut safety corners in drilling the well, violating federal regulations in the process;

—After completing the drilling, BP rushed to close the well, making many mistakes in the process;

—BP ignored final test results showing that the well had been improperly plugged.

Although BP should have done better, the same can be said of the federal government, which regulates offshore drilling. The

Interior Department, the primary agency responsible for oversight of the oil industry, simply was not equipped for the job, politically or practically.

THE ENVIRONMENTAL AND HUMAN IMPACT

The BP well blowout took a terrible toll on the environment. Flowing profusely and for great distances, the oil had devastating ecological effects on- and offshore. Ocean currents took the oil hundreds of miles away from the well site, and over a huge surface area of water. Ocean oil slicks reached more than 43,300 square miles, comparable to the total land area of Louisiana. Oil was also found on more than 400 square miles of the sea floor.[5]

Once it surfaced, oil worked its way in one direction to Texas, and in the other direction to Florida, hitting a total of some 1,300 miles of shoreline in the five Gulf states (Alabama, Florida, Louisiana, Mississippi, and Texas). Louisiana incurred the worst damage.[6]

Beaches and wetlands were contaminated with oil, as was wildlife. The effect on living organisms was toxic and often deadly, with estimates of dead animals ranging in the tens of thousands. Containment and cleanup efforts caused further ("collateral") damage. For example, the decision to use chemical dispersants involved a balance between breaking up the oil in the ocean and risking harm from exposing humans and marine wildlife to the chemicals. Similarly, boats entering wetlands to clean up the oil and retrieve boom that had flowed into them caused harm to the fragile marshes themselves.

The impact of the disaster on humans also was devastating. The fear of economic ruin was very real to hundreds of thousands of Gulf residents, many of whom were still recovering from the effects of Hurricane Katrina in 2005. As the oil gushed out into the water and up on the beaches, fishing and tourism suffered in a part of the country that relies heavily on both. The federal government closed down large areas of the Gulf waters to fishing. Tourists cancelled their trips to the region, not just to beach areas that had been hit with oil, but also to areas they thought might be hit. Given their lost income, Gulf residents were fearful about how they would pay rent; make payments on their homes, cars, or boats; and put food on the table for their families. Residents also feared the effects of the oil—and the tons of chemical dispersants used on it—on their health and the health of their children.

Clean-up workers especially had reason to be concerned about their health. Who were these "first responders?" Just about anyone who was willing to help. Offshore, they included fishermen who had been idled by the oil. Onshore, they included people from a variety of occupations who lost work due to the oil, but also many others, including the unemployed, homeless, and reportedly even inmates. Many were poor. Many were minorities.

By any measure, the response to the disaster was a major effort, involving nearly 50,000 workers at its peak. Many of the first responders were poorly trained, protected, and treated. Some became sick from exposure to the oil, the chemical dispersants, or the heat.

WELL CONTROL AND DAMAGE CONTROL IN THE GULF

There were two immediate tasks after the oil started spewing into the Gulf. First, stopping the oil from coming out of the well. Second, containing the oil that did. The U.S. Coast Guard was in charge of both missions, and it committed vast resources to them. The Coast Guard mobilized thousands of active duty and reserve personnel. At first, many Coast Guard personnel were put on standby in New Orleans area motels until their superiors figured out what they should do. Most had not been trained for this type of work. As a practical matter, neither the Coast Guard nor any other federal agency had the skills or equipment to respond to such a catastrophe. The staff of the Presidential Commission set up to study the disaster later observed: "When responders looked around in the government for specific expertise on well blow-outs, including in the military and in the scientific agencies, they found little to none."[7]

As a result, the federal government called in a little-known team of elite scientists known as the "Jasons," who often work on secret defense projects. President Obama turned to Steven Chu, his Nobel Prize–winning secretary of energy, to oversee the scientific effort. These acknowledged geniuses probably could solve just about any problem given enough time—even outside their normal areas of expertise—but time was one thing they did not have.

Although technically in charge, out of necessity, the federal government wound up relying heavily on BP. The awkwardness of this symbiotic relationship was on full display at an April 29 White House press conference when a Coast Guard admiral called BP a "partner." She was quickly corrected by Secretary of Homeland Security Janet Napolitano, who interrupted to say, "They are not our partner."[8]

In the end, the two entities did wind up working closely together on a number of fronts, ranging from technical issues to public relations. They established a Joint Information Center at the Unified Area Command Post in Robert, Louisiana, about 50 miles north of New Orleans across Lake Pontchartrain. There, they shared space in a Shell training facility leased by BP.

With all the amazing brainpower available, it still took nearly three months to come up with a workable solution to stop the oil flow.[9] It involved using the same type of cap that is usually applied to such wells—stacked on top of the existing one that hadn't worked. It took time to custom-build the cap, and there was fear that it might actually make things worse. Ultimately, the scientists decided that the risk was worth taking.

The cap worked, and the oil finally stopped flowing on July 15, 2010, 87 days after the blowout.

SPIN CONTROL IN WASHINGTON AND LONDON

The Mayday sounded in the Gulf was also heard in Washington and London. According to an official White House blog, President Obama was alerted that evening.[10] According to various reports, BP CEO Tony Hayward received the news at 7:24 A.M. London time (1:24 A.M. local time in the Gulf).[11]

The pressure on the president to stop the oil did not just strike close to home, but *in* his home. While he was shaving one morning, he recounted, his 10-year-old daughter Malia had asked, "Did you plug the hole yet, Daddy?[12]

The federal government and BP shared not only the goal of stemming the seemingly unending tide of oil, but also of repelling the impression that they were essentially impotent to do so. Sensitivities to image were especially high because both Hayward and President Obama were competing with the ghosts of prior Gulf tragedies. An explosion at a BP refinery in Texas City, Texas, that left 15 people dead in March 2005 was a major embarrassment for Lord John Browne, Hayward's predecessor. Hurricane Katrina, in August 2005, left more than 1,800 people dead and a permanent stain on the legacy of George W. Bush, Obama's predecessor.

The Obama administration was in an awkward political position from the start. Just three weeks before the BP well blew out in the Gulf, the White House had stunned environmentalists and its own allies in Congress by announcing that it was going to open up vast new swaths of American waters to offshore drilling, including areas off the Alaskan and East coasts. It staged the president's announcement at a military base, presumably to highlight how increased domestic oil production would improve the nation's security. Only a few months later, in his Oval Office address, President Obama portrayed BP's oil as the enemy.

The spin battle that summer produced a number of embarrassments. Initially, the administration grossly understated the amount of oil flowing from the well, and later grossly overstated the amount captured. The administration also misrepresented independent scientific support both for its decision to declare a moratorium on deepwater drilling and of a breakdown of the amount of oil captured. As Jane Lubchenco, head of the National Oceanic and Atmospheric Administration at the time, later put it, "The whole thing was a public relations nightmare."[13]

On August 14, 2010, a month after the well had been capped, the president took Malia swimming in the Gulf. The nation could infer that the water was safe again.

Meanwhile, BP was also struggling to put on its best face, and Tony Hayward was determined to be it. This turned out to be a big mistake. In the first of many gaffes that were both very revealing and ultimately would cost him his job, Hayward insisted early on that the amount of oil going into the Gulf was "tiny" relative to the size of the ocean.[14] A UCB Comedy spoof ("BP Coffee Spill") that went viral on the Internet ridicules that comment, as well as BP's inability to stop the flow.[15]

Hayward became a human punching bag when he testified before Congress. He was beaten up badly by Republicans and Democrats alike in both the House and the Senate. His repeated verbal fumbles brought disdain inside and outside the company. A high-level member of BP's public relations team described Hayward's public appearances as "a running sore."[16] President Obama said that he would have fired Hayward for his public statements. The CEO ultimately resigned in late July.

BP's own public relations nightmare turned into reality not just because of Hayward's misguided efforts but also because of Spillcam—cameras that BP had robotically placed at the wellhead. BP reportedly had initially resisted supplying footage from the cameras but ultimately caved in to a demand by Representative Edward Markey, chair of a House subcommittee investigating the disaster. After that, it was only a short matter of time before the subcommittee was relaying the footage to the press. Network television provided a worldwide audience with a live feed of the oil gushing into the Gulf, sometimes using a split screen for effect with other coverage of the disaster (such as when Tony Hayward testified before Congress).

The BP employees in the Robert Joint Command Center were watching CNN just like everyone else, and the BP public relations team soon came to scorn CNN (and CNN anchorman Anderson Cooper in particular). "The optics were terrible," said one of BP's top PR people.[17] By contrast, the optics were great for Markey, who used the footage in campaign ads when he ran successfully for the U.S. Senate from Massachusetts in 2013.

BP SURRENDERS AT THE WHITE HOUSE

The day after Obama's June 15 Oval Office address, BP's top brass came to the White House. Was the federal government picking on BP because it was a British company? Some thought so. "Make them stop calling us British Petroleum," BP's British marketing staff begged its American crisis management team.[18] The company had been rebranded simply "BP" in 2000, when a campaign was launched to identify those initials with "Beyond Petroleum" instead of "British Petroleum." British employees feared federal officials were intentionally using the outdated company name to stir up resentment against the company because it is British.

Whether inadvertent or intentional, such references likely served as less of a reminder that BP was a British company than Tony Hayward's British accent. Moreover, one prominent British magazine rejected outright the notion that the Americans were motivated by animosity towards the Brits.[19]

In a carefully staged series of meetings at the White House on June 16, 2010, top BP officials, including Hayward and Chairman of the Board Carl-Henric Svanberg, met with President Obama and other senior government officials. In what was likely a prede-termined outcome, BP agreed to set up a $20 billion trust fund to pay expenses related to the disaster. Svanberg was given a photo op with the president traditionally reserved for heads of state—seated in an armchair in front of the Oval Office fireplace.

THE PRIVATE CLAIMS

The $20 billion trust fund was used to pay for a broad range of expenses, including private claims. More than $6.2 billion was handed out in the first stage of the private claims program run for BP by a prominent Washington-based lawyer, Kenneth Feinberg. Feinberg brought to the task the gravitas, experience, and credibility of having been the administrator for the private claims in several major cases, including those resulting from the September 11, 2001, terrorist attacks against the United States. Nonetheless, he became unpopular with some of the Gulf claimants by insisting on the type of documentation that some of them could not produce given the cash economy in which they operated.[20]

Feinberg was eventually replaced by Patrick Juneau, a seasoned mediator from Louisiana. Juneau operated under a different set of guidelines, one that BP and the private parties had agreed upon in a lengthy court-approved settlement, and was overseen by the court. The new rules were less stringent than Feinberg's in some respects.

In the summer of 2013, BP started to push back hard against the court-administered claims process, fighting aggressively in court first to get the entire settlement thrown out and then to get Juneau removed. BP's hard-press offense offended the judge overseeing the case and ultimately failed. By April 2016 Juneau's awards totaled nearly $7.4 billion. Together with Feinberg's awards, then, BP was responsible for paying some $13.5 billion in private claims in the six years following the disaster.

No doubt some people tried to take advantage of BP's deep pocket by filing false claims, but it was the administrators' job to catch them. Some of the fraudulent claims, whether relatively petty or serious, wound up in the hands of the Justice Department for prosecution. A handful of employees at the Pensacola Beach, Florida, branch of the Hooters restaurant chain known for scantily clad waitresses, exaggerated the amount of their lost wages or helped others to exaggerate theirs.[21] Another Hooters employee claimed to have been let go from the Pensacola Beach location because of the disaster when he had really been fired from an inland branch for unrelated reasons.[22] A creative Michigan resident said he was stranded for 15 days in the Gulf on a boat fouled by BP's oil. He was never even there.[23]

The Justice Department reported that it prosecuted more than 300 individuals for BP claims fraud. Seventy-five defendants received prison sentences, some of them very substantial. The sentences of the Hooters employees ranged from six to 33 months. The Michigan offender was sentenced to 15 years in prison—the equivalent (coincidentally) of one year for each day he said he was adrift on his boat in the Gulf of Mexico.

THE UNITED STATES V. BP

The federal government brought criminal and civil cases against BP as a result of the accident. The company avoided a trial in the criminal case by agreeing to plead guilty to manslaughter, obstructing Congress, and environmental crimes—and to pay a record $4 billion.[24] In accepting the plea in January 2013, Chief U.S. District Judge Sarah Vance noted the need for the record payment "to protect the public from future misconduct by BP."[25]

The civil case against BP went to trial on January 20, 2013, in New Orleans. The line to watch the proceedings had started to form the night before, during a torrential thunderstorm, with placeholders making as much as $100 each. Just as soon as the courthouse opened at 7:30 A.M., the lawyers, reporters, and observers began filing in. The courtroom quickly filled to the brim. Lawyers even sat in the jury box since it was not a jury case.

Extra courtrooms had been set aside for video feeds to the overflow crowd.

A different federal district judge, Carl Barbier, presided over the civil trial. A native of Louisiana, with a touch of a Cajun accent, he had been handpicked by a special panel of federal judges to preside over most of the BP disaster civil litigation.

The stakes were high. BP was facing a maximum civil penalty of more than $16 billion. Ordinarily, under the Clean Water Act, any penalty would go in its entirety to the U.S. Treasury. But this was no ordinary case, and Congress had passed a law, sponsored by Louisiana senator Mary Landrieu, allocating 80 percent of any civil penalty to Gulf remediation. The law provided extra incentive for Gulf residents to hope that Judge Barbier would throw the book at BP.

The trial was split into three phases that continued over the course of two years, with lengthy breaks in between. In a nutshell, BP lost big in the first phase; the second phase was a draw; and BP settled before the judge issued an opinion in the third phase.

In the first phase, the judge found that BP had been "grossly negligent" in its conduct leading up to the blowout. This meant that the maximum civil penalty allowed by law would be $4,300 per barrel, nearly four times what it would be if BP had been only negligent. In the second phase, BP got a break. The judge found that more than 3 million barrels of oil had been discharged from the well, about a million fewer barrels than the government alleged. Thus, the maximum civil penalty would be $13.7 billion.

In the third phase, finally concluded in February 2015, the government and BP debated the amount of the penalty. Predictably, the government argued the penalty should be at the top end of the scale, whereas BP argued for the low end. Part of BP's argument was that it had suffered a drop in revenue due to lower oil prices.

The phase three decision was expected in early summer 2015. But just before the July 4 holiday weekend, BP suddenly announced a settlement in principle. BP would pay roughly a record $20 billion, over time, to finally put to rest all of its remaining civil liability for the disaster. This included a $5.5 billion civil penalty, $8 billion in natural resource damages, and nearly $6 billion in state and local economic damages.

THE COST OF THE ACCIDENT

What ultimately will be the total cost of the BP disaster? Just before BP announced major budget cuts and layoffs in early 2015 in the wake of the decline in oil prices, Hayward's successor, CEO Bob Dudley, attended the World Economic Forum. This annual extravaganza of corporate and world leaders takes place in the posh ski resort town of Davos, Switzerland.

As a "strategic partner" of the Forum, the highest-level membership reserved for a "select group of 100 leading global companies," BP gets five invites.[26] The cost of strategic partnership: 600,000 Swiss francs (roughly $685,000 at the time of the 2015 Forum).[27]

At Davos, Dudley was quoted by the BBC as saying that BP's accounting charge of more than $43 billion at the time to cover the costs of the disaster exceeded the costs of Hurricane Katrina.[28] The BBC speculated that Dudley was counting only insured hurricane losses, given that most estimates of total Katrina damage exceed $100 billion.[29]

By mid-2016 BP had increased the cost accounting charge to nearly $62 billion.[30] The true cost of the disaster, when the additional costs to society are included, possibly could be closer to the total cost of Katrina. Based on available information, BP is responsible for the world's costliest manmade corporate disaster.[31]

THE "MOTHER OF ALL RESTORATIONS"

In 2014 a panel of journalists covering environmental issues addressed a Washington, D.C., audience at the Woodrow Wilson Center in the Ronald Reagan Building. A questioner inquired about the long-term environmental remediation of the Gulf of Mexico necessitated by the BP disaster, a process known as restoration. He coined the term "The Mother of All Restorations."

The official federal and state government natural resource damage assessment for the restoration was finally completed in 2016. Under the civil and criminal settlements, more than $15 billion will be available to fund it. Projects are already under way, many of them very expensive ones. For example, the cost to restore one 7.5-mile stretch of critical beach and dune habitat on a Louisiana barrier island was estimated at more than $113 million.

LESSONS LEARNED

The BP disaster provides important lessons in corporate leadership, risk management, and sustainability—economic, social, and environmental—as well as in public policy.

BP's leadership put undue emphasis on profits and insufficient weight on compliance with federal laws and regulations. These leadership priorities—especially misguided given BP's past record of environmental violations—filtered down to the operational level. BP's leadership should have made legal compliance a higher priority.

BP should have invested in a state-of-the-art risk management program, especially given earlier deadly and damaging incidents. Instead, it failed to heed systemic warnings signals, just as it failed to heed specific danger signs just before the well blowout.

BP lacked progressive sustainability practices. It failed to properly take into account the social, environmental, and economic ramifications of its actions, causing immense harm on all three fronts.

In light of changing circumstances in the energy field, it is time for the United States to more closely examine how much offshore oil drilling it permits, especially from deepwater wells. The United States also needs more stringent regulations and enforcement to ensure that there are no more disasters.

2
THE BLOWOUT

Planes don't just fall out of the sky, and mobile oil rigs don't just explode—unless something goes very wrong. There was a massive explosion on board the *Deepwater Horizon* the night of April 20, 2010, as its crew worked to seal the 18,000-feet-deep oil well that sat below it. Something had gone very wrong. The well had blown out.

At the time, BP was at the end of what is known in the oil business as the exploration phase at the well. Drilling was complete, so the well would be closed temporarily and the rig detached from it. The production phase—when the oil would be pumped—would come later.

The well was called "Macondo," the name of the fictional Colombian village in the works of novelist Gabriel Garcia Marquez. As the well was being drilled, the oil workers on the rig also began to call it the "well from hell" because of multiple complications that the drillers encountered. Those complications were foreseeable given the geologic environment and the extreme depth of the well. Scuba divers who are familiar with the water pressure they encounter on a 100-foot dive can only imagine the pressure at 18,000 feet.

When the pressure at a well is not properly controlled, a blowout can occur, causing oil, natural gas, and/or other fluid to blast out. That's what happened at the Macondo well on April 20, 2010. The blowout triggered an explosion and fire aboard the *Deepwater Horizon.*

There were 126 people aboard the rig at the time of the blowout. Eleven men died either from the explosion or the inferno that followed. None of those who died actually worked for BP. Nine worked for Transocean, the owner of the *Deepwater Horizon.* BP leased the rig from Transocean, a major Swiss company that owns and operates such rigs all over the world. The other two dead were subcontractors of Schlumberger, the mammoth French global drilling services company.

The two youngest who perished, aged 22 and 24, were set to leave the rig the very next day. At least 17 others were seriously injured.

Two days after the initial explosion, the rig sank, and the pipe connecting it to the well was severed. At that point, massive amounts of oil began shooting out of the well, blasting in microscopic droplets, as if coming from an aerosol can. Some of the oil reached the water's surface and was drawn by currents across the Gulf. The rest remained in what is known as the water column, the area between the ocean surface and floor.

Notwithstanding the efforts of some of the best and brightest scientists in the country, both in the oil industry and the government, the well was not plugged for 87 days. During that time, nearly 3.2 million barrels of oil drained into the Gulf, at an average rate of nearly 37,000 barrels/day before it finally stopped. At 42 gallons to a barrel, this was the equivalent of about 1.5 million gallons per day. The effects on the Gulf, its people, and the environment were devastating.

A "DISASTER," NOT A "SPILL"

What happened in the Gulf beginning in April 2010 was a disaster, not a mere spill. "Spill" is a commonly used term of art used to describe an incident when oil is discharged into water or on land. But this was no ordinary spill in any sense of the word.

"Spill" is far too mild a term in terms of the quantity of oil that was released. A spill is much more like a minor leak such as milk in the kitchen, beer in the bar, or fuel in the gas station. The release of 3.2 million gallons of oil into the ocean is not a minor leak.

"Spill" is also far too mild in terms of fault. The well exploded because BP had been grossly negligent in a number of ways: drilling it, monitoring it, and sealing it. A spill generally comes about because of mere carelessness. BP's conduct was much more than just everyday carelessness.

"Spill" is also far too mild in terms of financial consequences. As noted, the costs to BP and society could approach the costs of Hurricane Katrina. A spill just doesn't cost that much.

Perhaps most important, "spill" is far too mild in terms of loss of life. Fred Hartley, the president of Union Oil at the time of the monumental 1969 oil rig blowout off the coast of Santa Barbara, California, said: "I don't like to call it a disaster because there has been no loss of human life."[1] Conversely, since 11 people died, there must have been a disaster.

WHAT WENT WRONG AT MACONDO?

According to the court's ruling in the consolidated civil suits brought by the federal government and private plaintiffs against BP, a series of blunders in the drilling and sealing of the well resulted in the blowout.[2] Three of these blunders are particularly striking.

First, the well had been improperly drilled. As the drilling neared completion at over 18,260 feet, BP decided to go still deeper. The decision to go the last 100 feet, according to a drilling expert who testified at trial, was "one of the most dangerous things [he] had ever seen in [his] 20 years experience" and was "totally unsafe."[3] Based on that testimony, the court concluded that the additional depth "left the wellbore in an extremely fragile condition." The court also found that "BP's decision to drill the final 100 feet was the initial link in a chain that concluded with the blowout, explosion, and oil spill" and "was motivated by profit."[4]

Second, in a fatal combination of haste, recklessness, and poor judgment, BP made major mistakes in plugging the well once the drilling was complete. BP planned to seal the well temporarily with cement so the *Deepwater Horizon* could move to its next job. Another rig eventually would replace it for the lucrative oil production phase. Although BP retained Halliburton, a major Texas energy services firm, to assist with the cement effort, BP maintained control over the project.

At the point that drilling was complete on April 9, 2010, BP was $60 million over budget and 54 days behind on the well. And the cost figure went up by $1 million each day that the *Deepwater Horizon* remained at the well.[5] That was perhaps a drop in the bucket for one of the world's richest companies. Its profit in 2009 was $16.75 billion. Nevertheless, there was pressure to get the job done.[6]

In the process, the wrong methods were used to ensure that the cement was properly applied. As a result, the cement casing essentially had holes in it, through which oil, other fluids, and natural gas could penetrate. It was only a matter of time before they did, and the well blew out.

BP repeatedly ignored red flags raised about engineering issues while the cement job was in progress. For example, an e-mail exchange between BP managers on a Saturday morning just three days before the explosion suggests that the managers were more concerned about finishing the cement job than addressing problems with it. Both managers were on the operational—not engineering—side of the house. One complained to the other about problems identified by an engineer on the rig. "Paranoia . . . from engineering leadership," he wrote. "What is the extent of my authority?" he asked, seemingly suggesting that he wanted to overrule the engineer. "I've got to go to dance practice in a few minutes," responded the more senior manager. "Let's talk this afternoon. . . . We're dancing to the Village People."[7] They did not speak again that afternoon.

The third, and perhaps most critical, blunder was BP's disregard of all-important test results for the well seal. These tests clearly showed that at 7:55 P.M., just two hours before the explosion, there was pressure on the drill pipe (one of the lines connecting the well to the rig), indicating fluids were moving dangerously through the casing. As one expert would later testify at trial, this so-called negative pressure test was a "pass-fail" test, and BP had failed.[8] The test should have triggered a "high alert" status. Nevertheless, BP deemed the cement job a success.

Those very test results were reported the night of the explosion in an 8:52 P.M. phone call from Don Vidrine, one of the BP supervisors on the rig, to Mark Hafle, the BP senior engineer in Houston:

> Vidrine told Mark that the crew had zero pressure on the kill line [another one of the high-pressure lines], but that they still had pressure on the drill pipe. Mark said he told Don that you can't have pressure on the drill pipe and zero pressure on the kill line in a test that's properly lined up.[9]

Although this quotation comes from BP investigators' interview notes, BP's subsequent Accident Investigation Report omitted any reference to this critical phone call, concluding that "the investigation team has found no evidence that the rig crew or well site leaders consulted anyone outside their team about the pressure abnormality."[10] The court found this statement to be "patently false" but stopped short of calling it a cover-up.[11]

Earlier in the process, the same BP senior engineer in Houston had thought that the cement design was so poor that it would result in a "shittie" job.[12] Nonetheless, despite being told about the inconsistent test results, he did not order further troubleshooting and testing. The court found this lack of follow-up to be "inexplicable."[13]

After the accident, one of the BP supervisors on the rig emailed BP's vice president of drilling and completions for the Gulf of Mexico, attributing the misinterpretation of the negative pressure test to what the supervisor called the "bladder effect." The vice president, who through sheer coincidence had been one of the VIPs on the rig that night to hand out the safety award, responded to the e-mail with lines and lines of questions marks—more than 500 of them. Neither the vice president nor any of the witnesses who testified at trial had ever heard of "the bladder effect" as a viable explanation for the negative pressure test failure.

ABANDONING SHIP

Within an hour of the critical phone call about the failed test, the well blew out, and fluids and gas started rising rapidly in the pipe. At 9:49 P.M. the gas ignited on the *Deepwater Horizon*, causing explosions and starting a massive fire. Four minutes later, bridge officer Andrea Fleytas sent the first Mayday.

There was chaos on board. In the aftermath, differing accounts emerged of what happened next. Chief Mate David Young testified that he had hit the General Alarm button once he saw fire, but by then the power had gone off, and he wasn't sure if an alarm had actually sounded. The public address system was still functioning and was used to alert the crew to the situation. Young later physically took the captain outside the bridge to show him the fire to convince him to abandon ship.

There was also considerable confusion on the bridge about whether to detach the rig from the well using the "Emergency Detachment System." It was unclear to the crew, including even the captain, which crew member had the authority to issue the order to activate the system. The captain thought that because the rig was not actually a moving vessel at the time, he needed to get permission from the person in charge of the drilling operation.

At least part of the confusion can be attributed to lack of adequate training. For example, the captain had not received required "major emergency management training." Moreover, training at the time did not include simulators such as those used to prepare airplane pilots to cope with emergencies.

Although it was apparently too late by the time of the blowout to save lives by detaching the rig from the well, it is possible that the explosions would not have been as severe and that the subsequent fire might have been extinguished.

At about 10:20 P.M. the surviving crew members abandoned ship. Most went into lifeboats. Four jumped directly into the sea some 75 feet below. Fortunately for the crew, other ships were nearby. They included the *Bankston*, which had been providing supplies to the *Deepwater Horizon*. Just minutes before the explosion, when the crew of the *Deepwater Horizon* realized that the well had blown out, they warned the *Bankston* to move away. The supply vessel was able to pick up the survivors from life rafts, as well as to pluck a handful from the ocean itself. The U.S. Coast Guard's search and rescue mission did not find any others.

THE COAST GUARD RESPONSE

After the rig sank on April 22, the U.S. Coast Guard, which has primary responsibility for responding to oil discharges, brought huge resources to bear in the Gulf. Some 8,000 personnel rotated through the Gulf in the year following the accident, with 3,000 there at the peak of the response. Slightly more than half were active duty personnel pulled from their normal duties, leaving large voids behind that the Coast Guard deemed "a calculated risk to their regular missions."[14] The remainder of the personnel were activated from the Coast Guard reserves, some from as far away as Guam. Vessels were brought in from around the country, meaning that they, too, were unavailable to respond to needs in their assigned sectors.

Notwithstanding the magnitude of the response, the Coast Guard was ill-prepared for the enormity of the task. Captain (now Rear Admiral) Meredith Austin likened the situation to having a "new major oil spill every day for 87 days."[15] The Coast Guard simply was not equipped to cope with anything like what it faced in the Gulf. Its on-scene coordinator later concluded "that significant improvements need to be made in preventative technology and requirements, mitigation technology and required capabilities, and oil spill response methods and readiness."[16]

Overwhelmed, the Coast Guard essentially turned over primary responsibility for the response to BP. The Coast Guard continued to contribute to the effort and to oversee the company's actions, as is typical in many oil pollution cases. Although they worked as cooperatively as possible, the awkward symbiotic relationship sometimes led to considerable disagreement, as well as the occasional shouting match.

BP'S RESPONSE

What became clear very quickly was that BP also was not prepared to cope with the aftermath of the disaster. The techniques BP initially used to try to stop the oil at the well source repeatedly met with failure.[17] The techniques that BP used to contain the oil that reached the surface had limited success and proved to be controversial.

Initially, BP pinned its hopes on the well's "blowout preventer." Then, BP tried to deploy a "containment dome." Next, BP attempted industry procedures known as "top kill" and "junk shot." All of these techniques failed, and the oil continued to pour out of the well until July 15. Then, it was finally plugged using a "capping stack." This was a smaller blowout preventer mounted on top of the failed one.

While working to stop the flow at the source, the company tried various methods to contain the oil. Large rings of absorbent material known as boom were placed into the water in an attempt to stop the oil from reaching shorelines. Although the boom had very limited effectiveness, its deployment at least created the appearance that something was being done. A Coast Guard reservist in the Gulf coined the expression "political boom" because it "didn't do anything and we knew it wasn't going to do anything."[18] It turned out the boom wasn't just a harmless public relations prop. Some of the boom flowed into marshes, which were susceptible to damage by boats that later came to find and retrieve it—no easy task.

Still, the boom was in hot demand, and BP couldn't seem to get enough of it. Garret Graves was Louisiana governor Bobby Jindal's point person for the disaster at the time, and continued to monitor its aftermath after being elected to the U.S. Congress. His impression of BP: "Here I was looking at this huge multinational, and they couldn't even get boom. . . . These guys couldn't seem to get themselves out of a wet paper bag. It was like watching a circus."[19]

An estimated 6 million gallons of oil were set on fire in what were described as "controlled burns." Conducted over a period of nine weeks, according to government scientists, the burning released 1.4 to 4.6 million pounds of black carbon (soot) into the atmosphere.[20] Physicians treating sick cleanup workers worried that some of their symptoms might have been caused by the pollution.

Responders also used a chemical dispersant—nearly 2 million gallons of it—in an attempt to break down the oil into smaller droplets. BP contractors dispensed most of it, both on the water's surface and into the ocean. Air Force C-130s reportedly flew 92 sorties, spraying nearly 150,000 gallons. Physicians were concerned about the dispersant's effect on human health, and environmentalists were concerned about the effect on aquatic life.

Eventually, the Environmental Protection Agency (EPA) ordered BP to cut back significantly on its use.

Ultimately, BP's handling of the crisis became the source of public mockery on the evening news, late-night television, and the internet. Thanks to robotically placed cameras at the wellhead, the public could watch a live-feed on network news of the oil incessantly streaming out. The public could also watch UCB Comedy's "BP Coffee Spill," a brilliant parody of some of the flawed techniques BP employed in its effort to stem the tide of oil. The clip culminates in a phone call to actor Kevin Costner to ask him for advice (which he had indeed volunteered). Posted on You Tube, the video had more than 13 million hits by late 2015.[21]

CORPORATE FAILURE AT BP

BP has often ranked in the top ten on the World Bank's list of the largest nonfinancial multinational corporations. How could such a prominent corporation allow such a terrible accident to happen, and then not be prepared to deal with it? The evidence suggests the company had put a premium on profits over safety.

Profits over Safety

Historical perspective provides some insight into BP's heavy focus on profits. The company that is now BP originated in the early twentieth century as the Anglo-Persian Oil Company, which evolved into the Anglo-Iranian Oil Company. Brookings scholar Kenneth Pollack reports that in 1949–50 Iran accounted for 76 percent of the company's total production, and was a key supplier of oil to the British, whose government was the majority shareholder.[22]

According to Pollack, the company earned itself a poor reputation in Iran: "They were determined to maximize immediate profits without regard for Iran or even for the ill will they were creating for themselves. The company lied and manipulated its books to underpay the Iranian government to the tune of billions of dollars. . . . The working conditions of [the company's] Iranian employees were unconscionable: they were paid 50 cents per day."[23]

Anglo-Iranian also has been linked to a checkered piece of mid-century Iranian history. In the early 1950s, tensions quickly built between Anglo-Iranian and its Iranian hosts, as the Iranian government nationalized all oil production, and the British plotted to overthrow it. The British enlisted the help of the CIA, which played the central role in engineering the 1953 coup that ousted the prime minister and put the Shah of Iran in power.[24] The Shah, widely regarded as a tyrant whom the U.S. government used as a strategic pawn during the Cold War, ruled until he was overthrown during the Iranian Revolution in 1979. The Clinton administration later acknowledged the American role in the coup.

Following the coup, in 1954, the company officially became British Petroleum. The British government began selling its shares in the late 1970s under Prime Minister Margaret Thatcher, and BP became completely private by the late 1980s. Privatization brought still-greater emphasis on profits, as well as expansion. Following a 1998 merger with Amoco, the company was known briefly as BP Amoco and became the largest producer of oil and gas in the United States. It then rebranded itself simply as "BP" in 2000.

Bob Bea, an emeritus Berkeley professor who began consulting for BP shortly thereafter, says he advised the company that it lacked the necessary safety culture. Bea says he pushed BP to adopt more rigorous process safety measures. Although the message got through to the midlevel managers, that's where it stopped, he said. Senior management simply refused to put in place changes that were necessary for managing risk at the substantial depths involved in deepwater drilling and forced into early retirement some of the managers advocating better risk management.[25]

Bea was positively emphatic that he had warned BP over and over again about its process safety problems, beginning as early as

2001, when he told them—in the blunt language of an oil man—"you are screwed."[26] Bea said that he told BP delegates at a 2007 conference in Normandy, France, "You still don't get it."[27] He ultimately blames BP management at all levels for the accident, saying that they put productivity and profits ahead of safety.[28]

A former BP executive and longtime accomplished industry insider who followed the accident closely after becoming an executive at another oil company, agrees that BP dropped the ball. The executive concedes that the blowout in the Gulf could have happened to any oil company: "We're an industry that needs to be fixed."[29] But, the executive qualifies, the chances were greater that such a calamity would happen to BP because it was so arrogant when it came to safety issues.

Still another industry veteran points out that with better risk management practices, rig workers (whether from BP or Transocean) would have had "Stop Work Authority" to challenge and address the incongruous negative pressure test results.[30] (Later, BP would announce that it had instituted such a practice.) Imperfect process safety may not have caused the accident, but state-of-the art risk management may have prevented it.

The Presidential Commission found that "BP's safety culture failed on the night April 20, 2011." "BP, Halliburton, and Transocean did not adequately identify or address risks of an accident—not in well design, cementing, or temporary abandonment [sealing] procedures."[31]

This theme was front-and-center during congressional investigations into the disaster in the Gulf. Representative (now senator from Massachusetts) Ed Markey, chair of a House subcommittee that investigated the disaster, linked the disaster to BP's culture and past record:

When the culture of a company favors risk-taking and cutting corners above other concerns, systemic failures like this oil spill disaster result without direct decisions being made or tradeoffs being considered. What is fully evident, from BP's pipeline spill in Alaska [2006] and the Texas city refinery disaster [2005], to the Deepwater Horizon well failure, is that BP has a long and sordid history of cutting costs and pushing the limits in search of higher profits.[32]

Response Planning

The bureaucratic paper trail provides strong clues to why BP's response to the disaster was such a fiasco. It simply did not prepare for the eventuality that occurred.

On March 19, 2008, BP purchased from the federal government for $34 million the lease rights to a nine-square-mile area off the coast of Louisiana anomalously named Mississippi Canyon.[33] After making the purchase, BP went through the process of submitting to federal regulators the necessary plans to obtain permission to drill a well in the area. BP's lengthy Initial Exploration Plan (EP) for the Macondo well was submitted in February 2009. In the section entitled "Blowout Scenario," BP wrote that "a scenario for a potential blowout of the well from which BP would expect to have the highest volume of liquid hydrocarbons is not required for the operations proposed in this EP."[34] In other words, the worst case scenario question was not applicable.

BP also submitted an Oil Spill Response Plan. It described the various species of wildlife that supposedly could be affected by an accident in the Gulf. In an indication that the Macondo plan was a cookie-cutter extract from another plan, some of the species identified in it (such as sea lions, sea otters, and walruses) exist not in the Gulf's warm waters but in frigid Alaskan waters.[35]

BP's cavalier approach to contingency planning clearly backfired. William Reilly, administrator of the Environmental Protection Agency (EPA) at the time of the 1989 *Exxon Valdez* accident and later co-chair of the Presidential Commission investigating the

BP disaster, said that he was "shocked" that BP was not better prepared than Exxon had been more than two decades earlier.[36] Reilly said that he will never forget how surprised he was when he called BP CEO Tony Hayward upon becoming commission co-chair and learned that BP had no subsea containment capability. "I was mystified," Reilly said, and "asked myself how that could be." Reilly also described Secretary of Energy Steven Chu as being "appalled" at BP's lack of monitoring equipment and technology.[37]

When the federal civil case against BP went to trial on February 25, 2013, it was in part the company's culture that was on trial.

As noted earlier, something has to go very wrong for a plane to fall out of the sky or for an oil rig to blow up. Flying could be more dangerous but is as safe as it is because both regulators and industry put a huge emphasis on safety. The airline industry has in place rigorous safety protocols and best practices. Pilots are trained, retrained, and tested, including in simulators, to handle emergencies. Flying in the United States now involves so little risk that there are fewer safer places to be than on a plane.

By contrast, although ultra-deepwater drilling is by definition hazardous, BP's protocols and best practices did not adequately correspond to the danger. The record suggests the company was simply going through the motions and playing its government regulator.

The BP disaster occurred because of risky behavior in an already high-risk venture.

3

THE ENVIRONMENTAL
AND HUMAN TOLL

The oil shooting out of BP's well spread mercilessly in the Gulf's waters and onto the surrounding shores. The oil in the ocean threatened sea life and the safety of seafood. The oil that washed ashore threatened more than 650 miles of Gulf coastal habitats that are especially rich in wildlife. Ten of thousands of animals were harmed or killed. Roughly half the oil discharged was thought to have remained below the surface after the well was sealed, prompting concern about the long-term effects on the marine environment, including on the food chain.

The short- and long-term effects on human health were also very real. Although the cleanup workers faced the most acute risk, the risk extended as well to people living along the Gulf coastline where the oil washed up. More than 37,000 medical benefit claims were filed, with over $20 million awarded. Because the disaster appears to have disproportionately affected minorities and the poor, questions arose about environmental justice—the intersection between environmental and civil rights issues.

THE ENVIRONMENTAL DAMAGE

The environmental damage was studied extensively by the federal and state government agencies designated as natural resource trustees. They concluded that the oil affected a wide swath of natural resources, including wildlife habitats such as sandy beaches and wetlands. Affected wildlife included fish, shellfish, birds, sea turtles, mammals, invertebrates, and plankton. Some of the affected marine life was protected under the Endangered Species Act.[1]

Living organisms suffered a range of toxic effects from exposure to the oil. These included death, disease, reduced growth, impaired reproduction, and life-threatening physiological impairments. The oil also had toxic effects on Gulf waters, marshes, and sediments that varied in degree and extent depending on concentration levels.[2]

Food Safety

Immediately following the blowout, fear arose that it was no longer safe to eat seafood coming out of the contaminated waters. The responsibility for assessing that danger was shared by the Food and Drug Administration (FDA), which oversees food safety, and the National Oceanic and Atmospheric Administration (NOAA), which oversees fisheries. They did not always see eye-to-eye on safety. Jane Lubchenco, NOAA administrator, said her agency had "knock down, drag out arguments with the FDA about what the standard should be."[3]

Along with other high-level administration officials, Lubchenco was personally dispatched to the Gulf early in the disaster. Louisiana state troopers tried to talk her out of a scheduled meeting with frustrated and worried fishermen in Plaquemines Parish, because they had been observed drinking in a bar near the meeting venue. The fishermen were upset because the oil forced the closing of large areas of the Gulf to fishing, a major industry there.

The Obama administration had gone to some lengths to recruit Lubchenco as the first female director of NOAA. John Podesta, who, with Rahm Emanuel, was heading up the selection of top administration officials after Barack Obama's win in 2008, called Lubchenco that December. Podesta asked her to fly to Chicago to meet the president-elect. A world-class scientist at Oregon State University, she was doing research at the time in Tasmania, Australia, nearly 10,000 miles from Chicago. She had no winter clothes with her, so she asked Podesta to put her up in Chicago near a department store. Properly outfitted after shopping, she met the president-elect and told him she would take the job. At a press conference the next day, Obama announced her selection along with the other leaders of his science team and formally nominated her for Senate confirmation later that month.

Lubchenco might have anticipated the culture change in going from academia to her first government position. Little did she know, however, that part of her job would be dealing with the nation's worst environmental disaster. She was very complimentary of NOAA's employees, some of whom were brought out of retirement to meet the human resource demands of the crisis. She was much less complimentary of the U.S. Congress, where, she said, "every member thinks they're your boss" and where she "had many many nasty hearings."[4]

By June 2, 2010, the peak of the fisheries closures, NOAA had shut down fishing in more than 36 percent of U.S. Gulf waters out of concern that the fish might be unsafe to eat.[5] During his third post-blowout trip to the Gulf, President Obama on June 4 did a photo op eating a shrimp, presumably to show seafood was safe in the rest of the Gulf. Similarly, Lubchenco made a point of saying, "I have confidence in our protocols and have enjoyed Gulf seafood each trip I've made to the region."[6]

The percentage of waters closed to fishing dwindled steadily through mid-November, when it reached near zero. But even after all the waters were officially open again to fishing, not everyone was following the president's example and eating Gulf shrimp, locally or elsewhere.

Wildlife

An abundance of wildlife lives in or transits the waters and lands of the Gulf. When BP's oil first washed ashore, it was on the barrier islands. As the weather turns cold in the southern hemisphere at the end of fall (the end of spring in the northern hemisphere), the birds head north, using the islands as resting places. Migrating birds were trapped and became soaked in oil when they landed. Images of oil-covered brown pelicans were depicted so frequently in television coverage that the bird became a symbol of the disaster.

The U.S. Fish and Wildlife Service initially reported that over 6,000 birds, over 600 turtles, and over 150 marine mammals (whales and dolphins) were found dead within a year of the accident.[7] Those figures left open the question of how many more had died but were not discovered, especially if they perished in the ocean.

Scientists have tried to fill in that informational gap through extrapolation and continue to study the long-term consequences of the oil on wildlife. In a 2015 report, the National Wildlife Federation (NWF) estimated the number of wildlife fatalities to be exponentially higher than earlier reported.[8] It put turtle deaths in 2010 at 27,000 to 65,000. NWF also calculated that some 12 percent of brown pelicans and 32 percent of laughing gills had died.[9] Dolphins off the Louisiana coast were reported to have died at four times historic rates in 2014, with "increasing evidence" connecting those deaths to the disaster.[10]

THE CLEANUP WORKERS

BP needed to mobilize a large number of people quickly to contain and clean up the oil. The company resorted largely to hiring contractors. Offshore, the workers included more than 10,000 fishermen and other boat workers who had been sidelined because of the disaster. Onshore, BP mobilized tens of thousands more workers to help with the cleanup, bringing the total number of cleanup workers to nearly 50,000 on a peak day.

The Boat Workers

BP launched what it called a Vessels of Opportunity program to recruit boat owners to help clean up and contain the offshore oil. The program's title was accurate, but paradoxical. As a result of government-imposed limitations on fishing, many of the local fishing boats were idled after the blowout. Many recreational fishing boats also remained in port. Although helping BP clean up its oil allowed many mariners to replace critically needed lost income, that "opportunity" only arose because of the disaster.

BP needed all the boats it could get for the cleanup, and paid well. A larger boat could make up to $3,000 per day. The pay was so good that Louisiana governor Bobby Jindal's office pushed for preference to be given to local fishermen. The company mustered an armada that it said totaled more than 3,000 fishing vessels. The reported cost of the program was $626 million.

Going out on the water to contain or collect the oil was also a potentially dangerous opportunity. Not oblivious to the risks, BP initially required the fishermen in the program to sign away their rights in case of harm in a one-sided 17-page contract. After seeing it, the president of the United Commercial Fisherman's Association compared BP's insistence on the waiver "to demanding that a person running into their own burning home sign a release limiting or giving up their claims against the arsonist who caused the fire."[11]

The president and association quickly filed suit in federal court, securing an emergency Sunday afternoon hearing on May 2, 2010, in federal court in New Orleans before Judge Ginger Berrigan. She also took issue with the contract. Two days later, BP came up with a more palatable version that the court approved.[12]

Training

Whether on boats or on shore, the workers faced health risks from the hazards of being exposed to the oil, the chemical agents used to disperse it, and the air pollution from the "controlled burns." Were the workers adequately informed of the risks and protected against them? Gina Solomon, then a physician with the Natural Resources Defense Council (and later deputy secretary for science and health at California's Environmental Protection Agency) was in the Gulf early in the disaster. She later said, "Taking a bunch of fishermen and quasi-training them to essentially become hazmat workers is dicey."[13]

The cleanup workers were supposed to fulfill an Occupational Safety and Health Administration (OSHA) requirement to undergo a minimal amount of training before becoming what BP called "Qualified Community Responders." The basic training course was four hours for onshore workers and four to eight hours for the boat workers, depending on the type of work. Supervisors were required to take a 40-hour hazardous waste operations ("HAZWOPER") course. The basic training reportedly was provided by BP-approved contractors without charge to the participants, but those taking the longer HAZWOPER training had to foot the bill themselves.

Numerous questions arose about the adequacy and effectiveness of the training, including whether it was sufficient to alert the workers to the risks and how to minimize them. Some workers reported not being given adequate information about potential health hazards. Others reported being told that they would be fired if they wore respirators and additional safety equipment.[14]

OSHA had a heavy presence in the Gulf during cleanup efforts, but it is not clear how effective it was. Judging by OSHA reports, some of the training contractors were just going through the motions to get the workers out the door and onto the beaches and boats. For example, OSHA was told "that some trainers are offering the

40-hour HAZWOPER training in significantly less than 40 hours, showing video presentations and offering only limited instruction."[15]

OSHA reported witnessing deficiencies in BP's safety and health program at several work sites and staging areas throughout the Gulf Coast region in late May, but it is not clear from available information that OSHA ever issued citations to the company or its contractors. Rather, for whatever reason, OSHA seems to have taken a light touch with BP during the Gulf disaster. By contrast, the federal government came down hard on one individual who, pretending to be a high-ranking OSHA official, offered certification of completion of a *six-day* hazardous materials training course after providing only *two hours* of training. The impersonator preyed on the Asian immigrant fishermen community in particular, using young bilingual speakers to lure participants through false enticements of subsequent employment in the cleanup effort, and collecting $150–$300 per class. She pled guilty to multiple felony charges and received a 57-month prison sentence.[16]

Housing

The cleanup workers who did not live in the areas where they were put to work required temporary housing. In some cases, the lodging reportedly provided was marginal.[17]

Some workers were housed on what became known as flotels. This housing consisted of steel rectangular boxes that resembled oversized shipping containers. They were stacked one on top of another on docked floating barges. The housing space per worker was reported to be 30 square feet, roughly a third less space than commonly found in a jail cell. Arguably, the space was sufficient because it technically was "offshore" and met U.S. Coast Guard standards. If the housing was considered land-based, OSHA standards applied and required more space.[18]

Other cleanup workers were housed in trailers. Ordinarily, trailers might have been considered an upgrade from the floating containers, but some of the trailers reportedly were the very same ones that had been supplied by the Federal Emergency Management Agency (FEMA) in the aftermath of Hurricane Katrina. These trailers gained notoriety when it was discovered that they contained formaldehyde (a potentially carcinogenic industrial chemical) at levels above federal safety standards. Despite the post-Katrina scandal and later FEMA warnings, more than 100,000 contaminated FEMA trailers reportedly had made their way onto the secondary market through public auctions.[19]

Given the known hazards from exposure to oil (discussed next), the cleanup workers should have been better informed, trained, and protected.

HEALTH EFFECTS

Early in the disaster, physicians and public health advocates expressed concern about the effects of human exposure to the oil. Their concerns focused on short-term high-concentration exposure (such as cleanup workers experienced); potentially longer term low-concentration exposure (such as permanent residents experienced); and psychological impacts (especially on the local communities).

The concern about potential health effects was grounded in decades of scientific research documenting the health effects of exposure to crude oil after accidental discharges. Clean-up workers typically complain of a variety of symptoms. These include headaches, sore eyes, sore throat, nausea, vomiting, dizziness, skin rashes, respiratory problems, and mental health issues. These symptoms are primarily due to exposure to the toxic polycylic aromatic hydrocarbons (PAHs) present in petroleum that make their way into the air, water, and soil.

Workers began reporting these types of health problems from the very start of the BP cleanup. A local physician who treated dozens of victims told CNN: "What's been really unique about it is that patients have come in with a severe amount of memory loss.

Very high blood pressure: blood pressures that are going sky high and then coming down to normal, and then blood sugar levels that are fluctuating. Lastly would be some pulmonary problems and some fairly serious [gastrointestinal] problems."[20]

In spite of ample anecdotal evidence of health problems, authorities were slow to begin and fund a systematic study. Some health experts lamented that seemingly more immediate attention was being paid to the ecological consequences of the disaster than to the human ones.[21]

Studies of the health effects of the BP disaster on both cleanup workers and local residents eventually did get under way and continue. A 2011 study found that the cleanup workers suffered acute health impacts mirroring those reported in previous major oil discharges.[22] The study recommended long-term surveillance for chronic adverse health effects including cancer, liver and kidney diseases, mental health disorders, and fetal alcohol spectrum disorders. A 2013 study of cleanup workers showed a variety of adverse health effects and focused in particular on potentially toxic effects on the liver.[23]

A 2011 epidemiological study of the psychological effects suffered by people in the Gulf region found that even residents who were not directly exposed to the oil had symptoms of anxiety and depression.[24] In a 2013 study, 40 percent of parents reported their children had experienced some type of adverse health effect since the disaster.[25] A 2015 study showed elevated levels of anxiety, depression, post-traumatic stress syndrome, and serious mental illness.[26]

The National Institute of Environmental Health Sciences, part of the National Institutes of Health, is conducting the GuLF Study, the largest ever study of the potential health effects associated with exposure to oil. The study plans to follow more than 30,000 members of the affected communities (cleanup workers and local residents) for ten years. Preliminary results, reported in 2014, revealed that cleanup workers were 30 percent more likely to suffer from depression or anxiety. Results reported in 2015 showed that the incidence of wheezing and coughing in cleanup workers was 20–30 percent higher than normal.[27]

Did BP pay sufficient attention to health issues in responding to the disaster? Early on, CEO Tony Hayward was quoted as suggesting that perhaps the cleanup workers had gotten sick from food poisoning.[28] Later, Richard Heron, the company's chief medical officer, responded to an article in *The Lancet*, the prestigious British medical journal, with the following statement:

> Based on extensive monitoring conducted by BP and the federal agencies, BP is not aware of any data showing worker or public exposures to oil constituents and dispersants at levels that would pose a health or safety concern. Indeed, results from more than 30,000 air monitoring samples . . . consistently showed that workers and members of the general public were not exposed to airborne concentrations of oil, oil constituents, or dispersants at levels above government mandated and recommended levels.[29]

Some experts and public interest groups remain skeptical, both of BP's assertion and its reliance on government standards and monitoring. "Situations like that in the Gulf bring into clear focus a problem we usually ignore," which is that OSHA standards are "horribly outdated," noted Linda Rae Murray, chief medical officer of the Cook County (Illinois) Department of Public Health and a past president of the American Public Health Association.[30]

Murray said no standards accurately account for factors that are important in real incidents, such as exposure to multiple toxic substances. Nor do they address the exposures and stress endured during extra-long crisis workdays that can go on for weeks or months. In those cases, "all our metrics go out the window."[31]

Some health care professionals also were concerned about the effects of exposure to the chemical dispersants used during the response. At times, it was difficult for caregivers to distinguish between symptoms caused by exposure to oil and the dispersant, since some of the symptoms overlap.

The Centers for Disease Control and Prevention (CDC) describes Corexit, the dispersant used by BP, as a combination of detergent and "low toxicity" solvent. CDC warned of potential health consequences to the workers handling and transporting the Corexit.[32] The warning made no mention, however, of potential exposure from aerial spraying in the same area where the Vessels of Opportunity boats were operating. CDC also assumed that any threat to the general public from Corexit being applied at the scene of the BP blowout would be unlikely "because of the strict guidelines that must be followed to utilize the dispersants."[33]

After first applying the Corexit only on the ocean surface, in mid-May BP sought and received authorization from federal authorities to also apply it underwater. The thinking behind this novel approach was that less dispersant would be required if it were to be applied subsurface, and that dispersant applied at depth would reduce the amount of oil that actually reached the surface. Of the nearly 2 million gallons of Corexit used to disperse the oil, roughly 700,000 gallons were applied subsurface.

EPA's authorization for the Corexit use was half-hearted, with Administrator Lisa Jackson hinting at choosing between the lesser of two evils.[34] Skeptics charged that dispersing the oil was geared less toward controlling its spread than controlling public relations—and that in any event toxic chemicals should not have been used. Although nontoxic dispersants existed, the technology had not yet been developed to scale up their production to the industrial levels necessary for such a large discharge.

The medical claims were settled in early 2012 during the same timeframe as the economic and property damage claims (discussed in chapter 5). As a result of the separate medical benefits settlement, cleanup workers and residents of specified geographic areas (beachfronts and wetlands) in 2010 whose health suffered as a result of the disaster are entitled to compensation from BP. To be eligible, they must have been diagnosed with a chronic condition prior to April 16, 2012, and have resided in the United States as of that date. They may receive payments of up to $60,700, as determined by a medical benefits claims administrator who reports to the New Orleans federal judge overseeing the BP civil litigation.

Under the settlement, BP also committed to establishing a Gulf Region Health Outreach Program, costing $105 million. The program comprises four divisions: a Primary Care Capacity Project, a Mental and Primary Care Capacity Project, an Environmental Health Capacity and Literacy Project, and a Community Health Workers Training Project. It is designed to serve the residents (especially the medically underserved) of 17 coastal counties and parishes in Alabama, Florida, Louisiana, and Mississippi.

ENVIRONMENTAL JUSTICE

Minority and poor communities are commonplace in the Gulf Coast region. There is evidence that a relatively high percentage of the cleanup workers came from these vulnerable populations. The disproportionate impact of the disaster on the poor and minorities raises important issues at the core of what is known as environmental justice. Gulf communities are all too familiar with these issues.[35]

As a movement, environmental justice is often first tied to major protests in 1982 against the dumping of toxic soil in a landfill in a black community in Warren County, North Carolina, and a 1983 General Accounting Office study that revealed that three-quarters of hazardous waste sites in eight southeastern states were situated in poor and black communities. (Now known as the Government Accountability Office, the GAO is an investigative arm of Congress.) The expression "Not in My Back Yard" (NIMBY) epitomizes the disparity between the powerful and the powerless in controlling environmental impacts in their communities.

The Gulf region historically has been linked to environmental justice issues. In 1990 Robert D. Bullard, considered by some to be the father of the environmental justice movement, wrote in his seminal book *Dumping in Dixie:*

> The entire Gulf Coast region, especially Mississippi, Alabama, Louisiana, and Texas, has been ravaged by "lax regulations and unbridled production." Polluting industries exploit the pro-growth and pro-jobs sentiment exhibited among the poor, working-class, and minority communities. Industries such as paper mills, waste disposal and treatment facilities, and chemical plants, searching for operation space, found these communities to be a logical choice for their expansion.[36]

A poignant example is "Cancer Alley," an 80-mile stretch of land that lies between New Orleans and Baton Rouge, Louisiana. It is home to oil refineries, chemical plants, and other industrial facilities; poor and minority residents; and illness. All in disproportionate numbers.[37]

The federal government first officially recognized the need to take action on environmental justice when President Clinton issued an Executive Order on Environmental Justice on February 11, 1994. Clinton ordered federal agencies with environmental responsibilities to take account of environmental justice issues in decisionmaking and established an interagency federal task force to address them. According to EPA, environmental justice "will be achieved when everyone enjoys the same degree of protection from environmental and health hazards and equal access to the decision-making process to have a healthy environment in which to live, learn, and work."[38]

Did the BP disaster have more of an impact on minority and low-income populations than on others? There is evidence that these vulnerable populations did indeed serve disproportionately as the "boots on the ground" (or the water) in the months after the blowout. CNN reported that some 4,500 of the onshore workers had been recruited though unemployment programs. Others were homeless (although it is difficult to know how many since homeless shelters in the Gulf received an influx of patrons from transients seeking work in the cleanup).[39] Some reportedly came from a parish jail work release program (and wore BP T-shirts) as well as a Louisiana state prison.[40]

These populations bore the brunt of the exposure to the oil and chemical dispersants, just as vulnerable populations historically have been overexposed to the nation's pollution. Given that vulnerable populations also generally have poorer-than-average health and access to health care, they may have been at higher-than-average risk of not receiving treatment for any adverse health effects they suffered.

Just as low-income populations with few employment options might welcome jobs in Cancer Alley, many of the BP cleanup workers likely did not have the luxury to turn down the work. CNN reported that some onshore workers were paid as much as $18 per hour, and supervisors as much as $32 per hour (although workers subcontracted to BP were said to make less).

BP's selection of the disposal sites for the voluminous amount of waste generated during the BP cleanup also raised questions (much of it was the containment boom of questionable effectiveness). Bullard wrote:

> Given the sad history of waste disposal in the southern United States, it should be no surprise to anyone that the BP disposal plan looks a lot like "Dumping in Dixie," and has become a core environmental justice concern, especially among low-income and people of color communities in the Gulf coast—communities whose residents have historically borne more than their fair share of solid waste landfills and hazardous waste facilities before and after natural and man-made disasters.[41]

Would BP have done a better job preventing, preparing for, or responding to the disaster had the backyards of more influential people been at stake? This broad question is difficult to answer. The evidence of disproportionate impact during the response was to some all too reminiscent of the effects of Hurricane Katrina not quite five years earlier. This time, however, the Gulf was struck not by a natural disaster, but by a manmade one.

4

BEHIND THE SCENES
IN WASHINGTON

The BP disaster became a top priority in Washington for the Obama administration and Congress. In the months following the blowout, the two branches of government struggled both practically and politically to address the crisis.

Practically speaking, managing the technical response to the seemingly endless flow of oil into the Gulf presented a difficult challenge for the federal government. It simply had limited capacity to deal with the problem. To its credit, early on the administration appointed a blue-ribbon commission to investigate the cause of the blowout and secured a $20 billion commitment from BP to pay expenses related to the disaster. The administration also imposed a short-term moratorium on deepwater drilling in the Gulf (albeit with an awkward rollout). To its discredit, the administration provided overly optimistic predictions and reports relating to the flow rate and capture of the oil. It just could not seem to get the numbers right.

Meanwhile, roughly a mile away from the White House up Pennsylvania Avenue on Capitol Hill, Congress was fully energized in an uncharacteristically bipartisan way immediately following the disaster. Various congressional committees held some 60 oversight hearings, covering topics that ranged from the liability cap that applies to oil discharges to measures for preventing future disasters. The committees heard both the heartfelt testimony of the families who had lost loved ones on the *Deepwater Horizon* and the evasive testimony of BP CEO Tony Hayward. Most important from a visibility standpoint, Congress produced Spillcam, the live-streaming footage of the oil pouring out of the well. Congress did not manage to pass any new laws that would give the public greater confidence in the federal government's ability to oversee offshore oil drilling.

THE ADMINISTRATION RESPONSE

On the morning of April 22, 2010, the *Deepwater Horizon* sank. At 6:00 P.M. that evening, President Obama convened a "Principals Meeting" in the Oval Office with the leaders of the key federal agencies responsible for handling the incident. Also present were senior White House aides, including Chief of Staff (now Chicago mayor) Rahm Emanuel and Assistant to the President for Energy and Climate Change ("Energy Czar") Carol Browner.

The risk of a major environmental catastrophe was looming. There was concern about the 700,000 gallons of diesel fuel on board the sunken vessel, and the even more substantial threat of oil being released from the well.

The president went around the room and asked the participants to address their agencies' oil release contingency plans, recalled Bob Perciasepe, EPA deputy administrator. Perciasepe was sitting in for EPA Administrator Lisa Jackson, who was in New York City observing Earth Day with Green for All, a Harlem-based environmental group. The president also had been in New York City earlier in the day to give an address on Wall Street reform at Cooper Union, a college in lower Manhattan. "The President and

First Lady's thoughts and prayers are with the family members and loved ones facing the tragic situation in the Gulf of Mexico," read the opening line of the brief White House statement following the meeting.[1]

At noon the next day, a Friday, the president and first lady left for a spring weekend in North Carolina's scenic Blue Ridge mountains. There, they might have looked forward to a relaxing couple of days enjoying barbeque, golf, tennis, and the spa. But bad news was to come. First, the Coast Guard's search for the eleven missing crewmen was suspended at 5 P.M. that afternoon. Then, on Saturday, came the first report that oil was spouting from the well into the Gulf.

Things continued to go downhill the following week. On April 28, BP reported an additional "breach" or leak, and estimates of the discharge of oil rose from 1,000 barrels a day to 5,000 barrels a day. The president was briefed on Air Force One as he flew back to Washington from a short trip to the Midwest that included touring an ethanol plant, and the administration started to ramp up its attention.

On April 29, the White House press briefing included an update on developments in the Gulf by Secretary of Homeland Security Janet Napolitano. In the major post-9/11 reorganization of the federal government, the new Department of Homeland Security had been granted lead authority over "domestic incident management."[2] The Coast Guard, the agency long charged with overseeing the nation's response to oil discharges in its waters, had been shifted from the Defense Department to Homeland Security. Thus, both lines of authority led to Napolitano.

On April 30, at what was scheduled to be Rose Garden remarks on the economy, the president opened with comments on the evolving crisis in the Gulf. He took the opportunity to reaffirm his continued belief in domestic oil production in the interest of national security. This somewhat defensive reference might have been part of a strategy designed to backstop his controversial decision less than three weeks earlier to open up more of America's waters to offshore drilling.

Over the April 30 weekend, Napolitano remained intensely involved in the federal government's response to the disaster. She appeared on four of the Sunday morning Washington talk shows, arriving for the first at CNN at 6:30 A.M. While in transit to the next studio, NBC, she wedged in a quick call from her limo with New York Mayor Michael Bloomberg to discuss the thwarted attempt to set off a car bomb in Times Square the evening before. Sunday afternoon, she had a teleconference with the Gulf state governors and two more conference calls with federal officials monitoring the disaster before her scheduled day ended at 6:30 P.M.

Over that action-packed weekend, Napolitano also announced the appointment of Admiral Thad Allen as the Coast Guard incident commander, the on-scene official in charge of response efforts. Allen, who had been brought in to take charge of the federal government's rescue efforts after Hurricane Katrina five years earlier, once again became the face of the government during a crisis in the Gulf.

That Sunday, the president made his first trip to the region "to inspect response operations firsthand."[3]

The Situation Room

With no progress having been made in stemming the steady tide of oil being unleashed into the Gulf, on May 10 the president convened another meeting of key cabinet and senior administration officials. This time, the meeting was held in the White House Situation Room. The president announced that Secretary of Energy Steven Chu would lead a team of top administration officials and government scientists to Houston that week "for an extensive dialogue with BP officials to continue to aggressively pursue potential solutions."[4]

Obama knew he needed someone with a solid scientific background to head up the administration's efforts to find a way to stop the oil—and not necessarily a politico. Chu was a natural choice. He was known to have the uncanny ability to think incisively about issues within or outside his field of expertise. This skill made him an ideal pick to spearhead the government's effort to seal the well.

The Situation Room became the venue for ongoing regular principals meetings. Typical of Washington power brokers, the cabinet members who were there initially squabbled about who was in charge. According to one person who was present, this led Carol Browner to shake her finger at them and remind them that they all worked for the same guy.[5] Although Napolitano officially chaired the meetings, Browner, as the president's direct senior aide, could trump the cabinet. It would be a long summer for Browner. She was a key administration player both in the battle for climate change legislation being waged on Capitol Hill and in the fight against the oil in the Gulf.

The deputy secretaries of the agencies involved also held regular "Deputies Meetings" in the Situation Room. Most of the operational decisions were made at the Incident Command Center in Robert, Louisiana, but any issues that could not be resolved on the ground were elevated to the Deputies Meetings.

The disaster had become an "all hands on deck" effort for government officials, not just in the Gulf, but also in Washington. It was an immense challenge for the various federal agencies, which sorely lacked the necessary expertise, resources, and training.

THE PRESIDENTIAL COMMISSION

A month into the disaster, oil was still gushing full force off the Louisiana coast and little was still known about what had caused the blowout. On May 22, a beautiful spring day in Washington,

President Obama stood on the White House lawn with former EPA Administrator William Reilly and former Senator Bob Graham. Obama announced that the two distinguished leaders would team up to co-chair a Presidential Commission to study the disaster. The commission was charged with determining what had gone wrong on the *Deepwater Horizon* and making recommendations about the future of offshore drilling. The president gave them just six months and a relatively limited budget of $15 million to do so.

Reilly was an ideal candidate to chair the commission. A well-respected moderate Republican environmentalist, he had been the president of the World Wildlife Fund before becoming EPA administrator in 1989 under President George H. W. Bush. A month after he took the reins at EPA, the *Exxon Valdez* tanker ran into a reef off the Alaskan coast, resulting in what had been considered the nation's worst environmental disaster prior to the BP blowout.

To avoid the appearance of a conflict, Reilly took a leave of absence from his position on the board of the Conoco Phillips Oil Company and took up residence in the Washington area to oversee the commission's efforts. Reilly's environmental expertise would eclipse Graham's during the course of the commission's hearings. But Graham, a highly accomplished and respected Democrat, later good-naturedly joked that Reilly had been telling him what to do since the 1992 Rio Earth Summit. (Graham was a member of the U.S. delegation that Reilly led to the summit.)

The historic Earth Summit had put climate change and world leaders on the stage together for the first time. Reilly worked hard there to fulfill his role, as he put it, as President Bush's "environmental conscience."[6] Despite lukewarm support from the Bush White House, the summit produced the landmark framework convention on climate change that eventually led to the 1997 Kyoto Protocol.

The Obama White House made another good pick for the commission's staff director: Richard Lazarus, one of the country's most distinguished environmental law professors. Then at Georgetown University, Lazarus went on to an endowed chair at Harvard. Unassuming in person, Lazarus had argued many environmental cases before the Supreme Court and had run moot courts (practice sessions) for other lawyers arguing environmental cases. Lazarus quickly put together a crack team of top-notch lawyers to assist in the investigation and drafting of staff working papers and a final commission report. They included former U.S. Supreme Court law clerks, considered the best and the brightest young lawyers around.

Although the commission was bipartisan by design, the seven commissioners got along relatively well, forging agreement fairly readily on most major issues relating to the disaster and offshore drilling generally, except exploration in the Arctic.[7] The commission's work was carefully researched and meticulously drafted—and decidedly critical of the government. So critical, in fact, that the administration started pushing back against what it perceived as bashing by the commission.

Reilly heard from EPA Administrator Lisa Jackson that the word coming out of the White House was that the commission was "not on message."[8] He heard similar complaints directly from Browner. NOAA Administrator Jane Lubchenco accused the commission of mischaracterizing her agency's work.[9]

Notwithstanding pushback from these senior officials, Reilly insisted that President Obama had told him to follow the facts wherever they led, and that the president was later "generous and complimentary."[10] It was not faint praise: the commission's report, issued in January 2011, provided excellent insight into the root causes of the disaster.

The commission found serious fault with BP and its federal regulator alike. It pointed to a "single overarching failure—a failure of management"—on BP's part.[11] It attributed the Mineral Management Service's "inexcusable" level of safety oversight to "decades of inadequate regulation, insufficient investment, and incomplete planning."[12] The commission's findings and recommendations will be discussed in further detail in this and later chapters.

THE INTERIOR DEPARTMENT AND THE MORATORIUM

May 2010 was also a busy month at the Interior Department, just a short walk from the White House. The president had given Secretary Ken Salazar until the end of the month to submit a report on increased safety measures for oil and gas exploration, and time was winding down.

Counselor to the Secretary Steve Black, an experienced energy and environmental lawyer who had held positions in the private and public sectors, was in charge of the team producing the report for the Interior Department. He was in close touch with Carol Browner's staff at the White House.

Black reached out to the National Academy of Engineering (NAE) for expert assistance. Seven prestigious NAE members agreed to help—on a voluntary basis. They promptly submitted a draft of their recommendations on May 10. The recommendations did not include a moratorium on oil drilling in the Gulf.

A moratorium was a touchy subject. On the one hand, the federal government had a responsibility to ensure that other deepwater wells did not explode. On the other, there were potentially significant costs associated with shutting down the thirty-three deepwater wells already in the process of being drilled in the Gulf, and suspending permitting for new ones.

On May 24 Black had a final conference call with the NAE experts to discuss Interior's report. It included Salazar's recommendation for a six-month moratorium on permits for new exploratory wells at a depth of 1,000 feet or greater and a temporary pause in all current drilling operations to allow for testing of the existing permitted wells.

On May 25 Salazar reportedly told Black that he had recommended the moratorium to the president. The president reportedly responded that he would sleep on it.[13]

If Obama slept on it that night, he did so in California, where he flew on May 25 to attend an evening fundraiser for Senator Barbara Boxer. The next day he spoke at Solyndra in Silicon Valley. The now infamous solar company filed for bankruptcy the following year, embarrassing the Energy Department, which had provided more than $500 million in loan guarantees. That morning, Black says, he got the approval from Salazar to include the moratorium in the executive summary of the report.[14]

On Thursday, May 27, the president was scheduled to hold a news conference on the disaster and then travel to the Gulf again the following day, ahead of Memorial Day Weekend. Interior was out of time. The report needed to be ready by the next morning.

So, it was a late Wednesday night for Black and his contact in Carol Browner's office at the White House. Black sent a version of the executive summary to the White House at 11:38 P.M., with the White House sending back an edited version at 2:13 A.M. The White House edited version had rearranged the wording in a way that implied that the NAE experts had endorsed the moratorium—which they had not.[15]

The president announced the moratorium at his news conference on May 27. No more offshore deepwater well drilling for the next six months. No new permits would be issued, and drilling would be suspended at the 33 deepwater wells already under way in the Gulf.

The final Interior Department report was also released that day—the version implying that the NAE experts supported the moratorium. The critical sentence of the report's executive summary read: "These recommendations contained in this report have been peer-reviewed by seven experts identified by the National Academy of Engineering."[16] "Peer review" means independent professional review and scrutiny. While the safety recommendations in the report had been subjected to peer review, that was not true of the moratorium.

Was the late-night change an intentionally orchestrated attempt to misrepresent or a careless, late-night drafting error? It's hard to say. Browner's office was said to be short-staffed, and the star of the office, Harvard law professor Jody Freeman, had left by then (after just over a year) to return to Cambridge mid-semester.[17]

The NAE experts were indignant when they found out that their names had been used to support the moratorium and promptly contacted Governor Bobby Jindal and the two U.S. senators from Louisiana, Mary Landrieu and David Vitter. Secretary Salazar and Deputy Secretary David Hayes soon apologized. The issue continued to haunt the Interior Department and triggered a congressional investigation by a key oversight committee once the Republicans took over the House of Representatives in the 2010 midterm election. The Interior Department relinquished key documents to the committee, but the White House stonewalled.

The moratorium became official with an order by Secretary Salazar on May 28, 2010, that became known as the "May Directive." It specifically barred both drilling under way at existing wells and the permitting and drilling of new wells, but did not affect production at wells already drilled.

Although environmentalists cheered the moratorium, it was much less popular in the Gulf, a region that thrives economically on the oil business. On June 7, 2010, Salazar's order was challenged in federal court in New Orleans by a group of offshore oil support contractors. The lawsuit was the first salvo in litigation brought by or against the federal government that related to the disaster.

THE PARTIAL BRITISH SURRENDER AT THE WHITE HOUSE

While the early legal skirmishing between the oil industry and the federal government over the moratorium played out in New Orleans, a major deal was in the works between BP and the White House.

On June 16 President Obama met with BP officials, including Chairman of the Board Carl-Henric Svanberg, to hash out final details. BP agreed to establish a $20 billion trust fund to pay expenses arising from the disaster, including private claims by victims in the Gulf. BP acknowledged that it needed to make good on such claims, but emphasized that it had also voluntarily established a separate $100 million fund to pay claims by oil rig workers who lost income as a result of the moratorium.[18]

In comments in the White House State Dining Room following the meeting, President Obama predicted that up to 90 percent of the oil coming out of the well in the coming days and weeks would be recovered. At the time the well was finally capped in mid-July, however, roughly 50 percent of the discharged oil remained in the subsea environment.

In comments made in the White House driveway (where television news cameras are permanently situated), Svanberg awkwardly referred to caring about "the small people" in the Gulf.[19]

Neither leader provided a solution to the ongoing geyser of oil erupting from BP's well. Nonetheless, the establishment of the trust fund was an important first step toward making the people of the Gulf whole for the economic losses they were suffering.

CONGRESS

On Capitol Hill, Congress swung into action after the blowout, not just in its elegant and ceremonial hearing rooms, but in the bowels of a House office building. Due to the diligent efforts of Representative Ed Markey and his committee staff, the public could watch on television or computers the live feed of the oil streaming from the wellhead on Spillcam.[20] The blurry image of the oil quickly became a symbol of the disaster. "America would have ignored it after five days, if it weren't for the video feed," Presidential Commission member Frances Beinecke later said.[21]

Markey's committee was a temporary one, originally set up by Democrats to help craft a major greenhouse gas reduction bill to address climate change. When the committee was created a few years earlier, office space was short, and the staff wound up in a room in a dark, dingy tunnel in the loading dock area of the Longworth House Office Building. Located just outside the fireproof doors of the building, some staff members feared that if there had been a fire, they would have been stuck behind the doors, with fatal consequences.

With the House having already passed a climate bill, the Select Committee's staff had time on its hands. Markey told them to focus on the BP disaster. Just before going into a closed-door hearing on May 4, Markey asked his staff for questions to pose to BP Vice President David Rainey. Aides pushed for him to ask about the size of the leak.

Rainey's answer to that very question—5,000 barrels per day— met with strong skepticism. Markey and his staff thought that Rainey was being deceptive about how much oil was flowing from the well in an effort to underplay the seriousness of the incident. They thought the number was much higher.

After Rainey's testimony, Markey's staff said they moved from "Defcon 3" to "Defcon 1." Defcon ("Defense Condition") 1 is a national security term used to describe maximum force readiness such as in the event of an impending nuclear attack. The staff was all but declaring war on BP's calculations.

The next shot was fired in a May 19, 2010, letter from Markey to BP America President Lamar McKay demanding that BP release footage from cameras placed robotically at the wellhead. The next day, BP provided Markey's committee with a link to some of the underwater footage.

The committee staff wanted to make the footage publicly available, but it had limited resources to do so. That night, they pulled an "all-nighter," cobbling together direct links from the BP rovers on the Gulf's ocean floor to Houston to an old Macintosh computer in the dungeon office.

The following day, Markey held a press conference announcing that Spillcam had gone live. Within hours, the link had gotten so many hits that the entire House website crashed. Within days, the Select Committee had posted links to all nine underwater cameras. Some major television news outlets began running live feeds, sometimes in split-screen mode. Viewers could see in real time the images of the oil spewing from the wellhead one mile below the surface of the ocean.

But every so often, the operators in Houston would have to reset the feed. When they did, the broadcast from the Mac computer providing the live feed would go to a black screen because the computer was password protected. Sometimes, in the middle of the night, Markey's press secretary would get a call from one of the networks saying the cameras were down. Markey's staffers would take turns getting out of bed and driving to the House office building. Down in the basement office, they would reenter the password into the Mac and revive the live feed to the underwater cameras so the world could continue watching. Scientists who had expertise in flow rates were also watching. They quickly determined that the rate was well in excess of 5,000 barrels a day.

Congress Bashes BP Top Brass

The powerful House Energy and Commerce Committee held oversight hearings covering the disaster. CEOs of rival oil companies, including Shell, ExxonMobil, Chevron, and ConocoPhillips, consistently bashed BP. So did members of Congress, whether they were from Gulf states or landlocked ones, and whether Republican or Democrat.

On June 15 when BP America President McKay testified, Representative Cliff Stearns, Republican from Florida, called upon him to quit. Republican colleague Anh "Joseph" Cao of Louisiana, a Vietnamese American, insinuated that McKay should go even further: "Well, in the Asian culture we do things differently. During the Samurai days we just give you a knife and ask you to commit harakiri."[22]

A few days later, a haggard and sheepish looking BP CEO Tony Hayward appeared before multiple committees in the House and Senate. One by one, members took a swing at him.

Bart Stupak, a Democrat, was chair of the House Energy and Commerce Committee's Subcommittee on Oversight and Investigations. He represented the northernmost district of Michigan, where residents have a reputation of being kind, salt-of-the earth Midwesterners.

Stupak was in no mood to be kind to Hayward. His anger visibly mounting, he asked Hayward what had happened that led BP's management to fail. Unsatisfied with the answer, Stupak pushed back hard: "Then what happened here? The head—the CEOs of the oil companies who were here before this committee Tuesday all said you did it wrong. They never would have done a well this way."[23]

Democrat Henry Waxman of California, chair of the House Energy and Commerce Committee, accused Hayward of failing to cooperate. When Hayward said he was distraught about the accident, Waxman wasn't interested: "I don't want to know if you're distraught," he said, and went on with his questions.[24]

Democrat Peter Welch of Vermont admonished Hayward that "your answer 65 times that you don't know to questions that were reasonably posed to you on both sides of the aisle erodes confidence; it doesn't inspire confidence."[25]

Hayward did have one stalwart defender: Joe Barton, the ranking Republican member of the House Energy and Commerce Committee. The shoot-from-the-hip congressman who represents

Houston, home of BP's U.S. headquarters, took it upon himself to defend the company. He apologized to Hayward for what he called a "shakedown." Barton was referring not to the bashing on Capitol Hill but the administration's leveraging of the $20 billion trust fund.

The Barton apology drew sharp criticism from many quarters, as well as news stories reporting more than $1.5 million in campaign contributions to him from the oil industry. Republican Jeff Miller from Florida called for Barton to resign as the ranking Republican on the committee. Rahm Emanuel also reacted sharply. Faced with such widespread criticism, Barton eventually retracted his apology.

The hearings made for entertaining theatrics. It remained to be seen whether Congress would pass legislation designed to prevent another tragedy.

Legislation Fails

Notwithstanding all of the hearings, Congress failed to pass legislation in 2010 either to ensure full reimbursement of cleanup costs or to improve regulation of offshore drilling.

After the blowout, the Senate focused quickly on a bill to lift the liability cap under the Oil Pollution Act that normally applies to oil discharges. Without such a change, BP's liability conceivably would be limited to $75 million in cleanup costs, unless it were found to be grossly negligent (an unknown at the time). A bill was introduced, but never passed. The hearing on the bill before the Senate Environment and Public Works Committee became so contentious that in a moment of exasperation, one moderate Democratic member of the committee vented privately to staff, "This is the most f_____ up committee in the entire Senate."[26]

With Democrats outnumbering Republicans on the committee, the bill eventually reached the Senate floor. It never came up for a vote, however, reportedly due to opposition from senators from oil and gas states who wanted the cap raised, but not lifted. Ultimately, the legislation would prove unnecessary since BP agreed to waive the cap.

Meanwhile, the House moved forward on a bill introduced before the blowout that would impose stiffer regulations on offshore drilling and overhaul the Minerals Management Service. The bill passed by just 16 votes, with 39 Democrats opposing it in the face of heavy lobbying by the oil industry. A less controversial bill expanding compensation rights for victims of offshore accidents also passed. Both pieces of legislation were sent to the Senate, where they sat until Congress adjourned in December. The end of the session meant the end of the bills. Republicans took over the House of Representatives in January 2011.

Why did the Senate not act in time? Typically, a crisis like the one in the Gulf would be considered agenda-setting in terms of stimulating passage of legislation. But the Senate had other pressing legislative priorities on its horizon when the well blew out on April 20, and, as it turned out, that work did not necessarily square with tightening offshore oil regulation.

In fact, the timing could not have been worse. Earlier on the very same day that the blowout occurred, key senators drafting the Senate version of the climate bill went to see Rahm Emanuel at the White House. The bill's sponsors and supporters had long labored to forge an acceptable compromise with business and environmental groups. Environmental groups cheered the bill's requirement that oil companies finally would be forced to account for the costs associated with the greenhouse gas emissions caused by their products.

Some Democrats saw a silver lining in the disaster. They believed it could help spur the effort to get the climate change bill passed that summer. The use of oil products is a major source of greenhouse gas emissions, after all, and oil drilling had resulted in the blowout. For this gamesmanship to play out successfully, however, the public would have to associate the disaster with Republicans.

This was a hard sell considering the support for offshore drilling by oil state Democrats—and the president's March announcement to expand it.

Others saw the BP disaster as an impediment to passage of the climate bill, rather than an opportunity. Reluctant Democrats from oil states conceivably could have been lulled into voting for the bill had it included incentives for the oil companies in their states. But such horse trading was politically infeasible while oil was still gushing into the Gulf.

Time was short. As a practical matter, because it was an election year, any legislation would need to be passed by the end of the summer. The climate bill timed out, and so did offshore oil regulation reform. The blowout in the Gulf ordinarily might have presented an agenda-setting opportunity, but it was a lost one.

5

THE CLAIMS AND THE CONTROVERSIES

"We will make it right," Tony Hayward proclaimed with his distinctive British accent, standing on a Gulf beach awash in BP oil. "We will pay all legitimate claims," he promised.

Many individuals and businesses had legitimate claims. Tourism and commercial fishing suffered significant losses as a result of the disaster. Both are major industries in the Gulf. Estimates put annual revenue from coastal Gulf tourism at $20 billion, and commercial fishing in Louisiana alone at $2.5 billion. Louisiana produces one-third of the nation's shrimp (another third comes from other Gulf waters) and two-thirds of its oysters.

Tourism declined heavily across the Gulf Coast. The closing of large areas of the Gulf to fishing also had severe financial impacts. Losses were suffered by business owners and employees alike. Fishermen, hotel workers, restaurant workers, and many others saw their incomes eliminated or reduced. Property was damaged or destroyed.

Private parties with financial losses had the option of filing individual lawsuits or joining a major class action lawsuit (involving many plaintiffs with common claims). They could also file claims without suing under a program initiated by BP and then continued under court supervision. Some 600,000 economic and property claims were submitted during the five-year period following the blowout.

True to Hayward's promise, BP's initial strategy was to take responsibility and make amends for financial losses. The establishment in June 2010 of the $20 billion trust fund to cover claims and other expenses was a positive step in that direction (resolution of the separate federal and state government legal cases against the company came much later). As the claims mounted, however, BP pushed back vigorously (but unsuccessfully) against the claims process.

Some individuals tried to take advantage of BP's deep pocket by filing fictitious claims or overstating legitimate ones. The Justice Department reported bringing criminal cases against more than 300 people who had filed false BP claims, winning 236 convictions.[1] Of these, 75 received federal prison sentences, one as high as 15 years.[2] By contrast, as discussed in the next chapter, only five individuals were ever prosecuted for crimes associated with the well explosion, 11 deaths, environmental devastation, and alleged cover-ups. None of them will serve time in prison.

THE MASTER OF DISASTERS AND THE CAJUN MASTER

The economic and property damage claims process (which was separate from the medical claims process established later) was initially overseen by an independent administrator hired by BP, but was later run by a court-supervised administrator. In general, BP much preferred the results under the first administrator, whereas the claimants much preferred them under the second.

The first claims administrator was Ken Feinberg, a well-known mediator and Washington power broker. Dubbed the "Master of Disasters" in an *American Bar Association Journal* article in 2011, Feinberg has the credibility of someone with deep experience in claims adjudication.[3] He has overseen some of the most high-profile and diverse claims in American history, including those filed by victims of chemical companies (Vietnam War veterans poisoned by Agent Orange), terrorism (9/11 and Boston Marathon), mass shootings (Virginia Tech), child abuse (Penn State), and automobile defects (GM).

Feinberg presided over the Gulf Coast Claims Facility that operated from August 2010 through March 2012. It processed some 220,000 claims and distributed more than $6.2 billion (just short of the $7 billion Feinberg had awarded under the September 11th Victim Compensation Fund).

Known for his independence, Feinberg explained that his job was to play Solomon—to judge the claims and get finality for the victims and BP.[4] As the claims came in—legitimate and illegitimate—Feinberg and his team weeded through them. Victims were paid for losses they could establish were connected to the disaster and document in terms of amount. In exchange, BP received a legal release from future claims. Since claims could be submitted without hiring lawyers, victims who chose to file one without legal representation did not have to share a percentage of their compensation under a contingency fee agreement common in such cases.

Feinberg tried to explain the claims process in town halls around the Gulf but commonly met with resistance. Although reportedly paid royally by BP for his role, Feinberg has a reputation as an impartial arbiter with an allegiance to fairness. But some of the Gulf victims and their lawyers did not necessarily see Feinberg in that light. Or, as Feinberg unabashedly put it himself: "The idea of a white, Yankee, Jewish Bostonian riding in and riding out with the blessing of the administration, funded by BP—there's a presumption before you even start that you're treading water."[5]

Feinberg was indeed a cultural misfit in Cajun country and other parts of the Gulf. More important, the rigor of his standardized documentation requirements jived poorly with the more relaxed, local laissez-faire approach to business practices and records. Whether a shrimp boat captain, a day laborer, or a waitress, not all claimants kept fastidious written records of their earnings.

The claims program Feinberg oversaw was terminated in March 2012, during the course of settlement negotiations over the private party class action litigation. Following a transition period, the Gulf Coast Claims Facility was permanently replaced by a new Court Supervised Claims Program.

The new claims program was headed by Patrick Juneau, a lawyer from Lafayette, Louisiana, in the heart of Cajun country. Juneau spent most of his career as a defense lawyer representing corporations in personal injury cases. He is a respected and experienced mediator, albeit on a much more local scale than Feinberg.

As of April 2016, the Court Supervised Claims Program reported receiving some 384,000 claims, of which 90 percent had been processed. Nearly 122,000 of the claims were paid, for a total of $7.4 billion.[6]

The new program operated under a different set of guidelines than the initial one. The new guidelines were included in the 1,000-plus-page agreement between BP and the private parties that settled the private party class action litigation. Approved in June 2012 by U.S. District Court Judge Carl Barbier, the legally binding consent decree included a significantly reduced threshold for claims adjudication proof that likely was debated over long hours at the negotiation table by the lawyers. Claimants that operated certain types of business in coastal areas most directly impacted by the oil were no longer required to provide documentation of "causation." In other words, claimants in this special category were exempt from demonstrating that BP's oil had actually caused their losses.

Causation is generally considered to be an important element of establishing a viable claim, so why would BP agree to such an uncommon and potentially expensive provision? Quite possibly,

BP was so interested in making a deal that it calculated that the limited exemption was not too much to pay for finality. Or so it may have seemed at the time.

BP PUSHES BACK

By the summer of 2013, however, the tide had turned. BP essentially did a double take and cried foul, arguing that it had never agreed to pay claims not tangibly related to the disaster. The company embarked on a two-prong attack on the causation exemption—both in the court of public opinion and in federal court.

BP launched a massive public relations campaign to drum up support for its position, beefing up its communications team in the process. The company announced that it would promote Geoff Morrell, its talented U.S. communications head, to the post of senior vice president. Bob Dudley, BP's new CEO, recognized Morrell for his "forward-leaning, assertive" voice.[7] This was a distinctly different Morrell voice than was heard in 2010, when, as Pentagon press secretary, he pledged to hold BP accountable for the blowout.[8] In 2011 Morrell had taken a swing in Washington's proverbial revolving door and joined BP.

BP's press release of Morrell's promotion hinted at what its public relations strategy would be:

As we continue to address the political and legal challenges we face in the United States, we have consistently conveyed BP's commitment to America and the contributions we make to the nation's economy and energy security," said BP America Chairman and President John Mingé. "I look to Geoff and his team to ensure that this message is delivered with absolute clarity day in and day out."[9]

To implement its public relations strategy, BP ran full-page ads in major national newspapers to assert how unfairly it was being treated:

Last year, we signed a settlement agreement to ensure that people who suffered losses from the accident would keep being paid. When we negotiated that agreement, we sat down in good faith with the goal of helping as many deserving people as possible. And when we signed it that's what we thought the agreement would do. Unfortunately, it's now being applied in a way that ignores the agreement's plain language, with enormous payments going to businesses that did not suffer any losses.[10]

The summer of 2013 also marked a turning point in BP's legal strategy. The company insisted in court that it had never made a deal waiving evidence of causation, and that Juneau was misinterpreting the settlement agreement.

Was BP being exploited, or was it simply looking for excuses to renege on the deal? Judge Barbier decided that BP was crying wolf. In an extraordinarily timed order issued on Christmas Eve in 2013, he wrote: "Frankly, it is surprising that the same counsel who . . . strenuously advocated for approval of the settlement now come to this Court and the Fifth Circuit [Court of Appeals] and contradict everything they have previously done or said on this issue."[11] The settlement stood.

BP repeatedly went over Judge Barbier's head, or, as they say in New Orleans, "across the street," to the U.S. Court of Appeals for the Fifth Circuit. Ultimately, BP even petitioned the U.S. Supreme Court. To aid in its effort, BP hired Ted Olson, the same formidable appellate lawyer who helped George W. Bush win the White House when the Florida ballot controversy erupted during the 2000 presidential election. BP lost soundly in the Court of Appeals, and the Supreme Court declined to hear the case.[12]

But BP wasn't through fighting back. In June 2014 the company retained yet another law firm, uber-litigators Williams and Connolly of Washington, D.C. They filed another suit to recover hundreds of millions of dollars that BP said were overpaid under the court-supervised program. The case involved a difference in interpretation of the accounting methodology of the very complex 1,000-plus-page settlement agreement. As its primary example, BP pointed to a seller of animals and animal skins that it said received a $14 million overpayment.

The claims cash register clicked away over the summer of 2014 while BP continued its massive public relations campaign, appealing to an American public that, like the courts, had grown weary of hearing about BP's perceived woes. The publicity ultimately reached a point that journalists were criticizing each other for being willful participants in a brainwashing attempt. A May 4, 2014, *60 Minutes* episode that was mostly sympathetic to BP's cause drew a sharp rebuke in the *Los Angeles Times*, with its business columnist writing: "The program did the company's work quite adequately, to its enduring shame."[13]

Whether in Feinberg's hands or Juneau's, the claims process did hit rough patches. Under Feinberg, claimants complained bitterly about how long it took to get their claims processed and how much documentation was required. Under Juneau, BP complained just as bitterly that claims were being overpaid.

In 2012 the Gulf Coast Claims Facility administered by Feinberg underwent an independent audit commissioned by the Justice Department that produced a mixed review. On the one hand, the audit found that nearly 7,300 claimants had been negatively affected by claims processing errors, resulting in underpayments of an estimated $64 million.[14] On the other, the audit recognized the achievement of processing a vast number of claims over a relatively short period of time.

In 2015 one of the facility's claims adjustors pled guilty to fraud for providing false documentation to assist claimants posing as commercial fishermen in exchange for a share of the take. She had netted $250,000, a portion of which she used to purchase a house with one of the fake fishermen. She was sentenced to probation.

The Court Supervised Claims Program administered by Juneau also underwent an independent examination, which was conducted by former FBI director (and federal judge) Louis Freeh amid allegations of misconduct by Juneau's staff. The allegations so concerned Judge Barbier that he swiftly summoned Freeh to New Orleans and commissioned him as a "Special Master" to conduct an investigation.

Known as a very thorough investigator, Freeh dug deep. In a September 2013 report, Freeh make a number of troubling findings about one of the attorneys in the claims administration office of the Court Supervised Claims Program. Prior to being employed by the program, the attorney represented a shrimp fisherman who subsequently filed a claim. Although the claims attorney referred the fisherman to another attorney prior to joining the claims administration office, while employed there the claims attorney expedited the payment of the other attorney's fees and received a referral fee. Not only did the claims attorney fail to disclose to Juneau his interest in the claim, but when confronted by him, also lied about it.[15]

To make matters worse, the claim was fraudulent. The claims program awarded the fisherman just over $350,000 for alleged lost fishing revenue based on tax returns he submitted to substantiate his past income. The tax returns for the same year that he filed with the IRS, however, said that he was unemployed with no income. Judge Barbier ordered the fisherman to return the money and the claims administration office attorney (who resigned shortly after the scandal surfaced) to return the referral fee.[16] The judge also banned the claims attorney and the claimant's individual lawyers from any further participation in the claims program and ordered that they be referred to their state bar disciplinary authorities.[17] The fisherman was subsequently prosecuted for fraud, pled guilty, and was sentenced to two years in prison.[18]

On the same day that Freeh issued his September report, Judge Barbier asked him to look at an additional conflict of interest situation. In the second case, Freeh found that the claims administration office appeals coordinator had passed internal information to a cousin working at a Louisiana law firm that represented a claimant, and had gone drinking with other claims administration office employees at a New Orleans bar owned by yet another claimant.[19] Such actions arguably constituted a violation of the general ethical canon that lawyers should avoid even the appearance of impropriety. Three more resignations ensued.

BP tried to capitalize on the problems with the Court Supervised Claims Program by asking Judge Barbier to remove Juneau. The judge refused, and BP appealed to the Fifth Circuit. BP's combative legal and public relations campaigns started to wind down in the spring of 2015, when BP withdrew that appeal. Then, in the summer of 2015, BP lost another appeal to the Fifth Circuit to have access to claim-specific information before claims were even adjudicated. In its ruling, the court noted the settlement agreement's preexisting framework for addressing fraud, former FBI director Freeh's role in fraud prevention and investigation, and the Justice Department's "high priority on promptly investigating and prosecuting all meritorious reports of fraud related to the oil spill and its aftermath."[20] In other words, sufficient safeguards were in place to identify and punish people who tried to take advantage of BP. The company's two-year campaign against the claims process was finally over.

THE FALSE CLAIMS

In addition to the types of irregularities discussed above, the claims process generated many instances of outright fraud. Charges were leveled against lawyers and claimants alike.

The BP disaster spawned a cottage industry of claims lawyers who saw a potential gold mine in representing disaster victims. Some specialized in personal injury cases, and others had little relevant experience.

One lawyer who seized the opportunity was Mikal Watts, an accomplished San Antonio, Texas, personal injury attorney. Also a major player in political circles, on his birthday in July 2012, Watts held a fundraiser at his home, attended by President Obama.

Watts became a target of both BP and the federal government. In 2013 BP sued Watts for his alleged role in bringing false claims against the company. The civil case was put on hold while the Justice Department conducted a lengthy criminal investigation into the matter. In September 2015 Justice indicted Watts and others on conspiracy, fraud, identity theft, and other charges.[21]

Watts was alleged to have falsely claimed that he represented more than 40,000 deck hands in an attempt to win a place on the class action Plaintiffs' Steering Committee (the most influential and well-compensated group of lawyers representing private plaintiffs).[22] To inflate his number of clients, Watts also was alleged to have filed claims using fake Social Security numbers.[23] The trial began in July 2016, with Ken Feinberg as the first witness for the government. Feinberg read aloud a letter to Watts saying, "It's hard to believe there are even 41,000 fishermen in the Gulf who would even file a claim."[24]

After representing himself at trial, Watts was acquitted in August 2016, as were two of his law firm associates and employees. The jury found no evidence that they were aware of the fraud.[25] Two other defendants who had provided the list of claimants to Watts were found guilty.[26]

There were hundreds of instances of individuals and businesses accused of filing fraudulent claims. Some large, some small, such claims often were for lost personal income or business revenues allegedly attributable to the disaster. Individuals submitted falsified pay stubs and other fake documents. Businesses used false documentation such as tax identification numbers, business permits, tax returns, and sales receipts.

From the start, the federal government took such fraud very seriously. Just one day after the Gulf Coast Claims Facility began taking claims, the Department of Justice announced it was prepared to crack down on any and all types of disaster-related fraud. The National Center for Disaster Fraud, already in place from Katrina days, was poised to handle such cases. To help coordinate Washington, D.C., and local federal prosecutors (known as the U.S. Attorney's offices), the department held a one-day law enforcement training and coordination conference on oil spill fraud issues. Law enforcement assistance was provided by both the FBI and the Secret Service. Some of the Gulf state local prosecutors also pursued the claims fraud.

A few cases are particularly noteworthy. The most brazen involved Duane Montgomery, a former Detroit mayoral and congressional candidate. Although he apparently still resided in Michigan at the time of the blowout, Montgomery submitted multiple false claims for damages, beginning in the summer of 2010. He pursued them aggressively, even when they were denied, through his summer 2013 conviction for fraud.[27]

In a particularly far-fetched tale, Montgomery claimed that as a result of "pollution monitoring" he conducted post-blowout in the Gulf of Mexico, tar balls from the oil spill destroyed his boat's engines, casting him adrift at sea for 15 days.[28] The appellate court affirming his conviction found that "he did not own a boat at the time, and certainly was not working in the Gulf during the oil spill."[29] Moreover, the court found that he had submitted fraudulent documentation to support his ill-fated fabrication. He had falsified tax returns. The insurance check for the so-called towing bill for his boat in the Gulf turned out to be reimbursement for a motorcycle that he reported stolen in Michigan. Photographs of boat engines he said were damaged by oil in the Gulf were actually for a sailboat in Delaware.

Montgomery chose to represent himself at his trial and in his subsequent appeals, failing at every stage. After the trial jury convicted him on three counts of fraud, the judge threw the book at him, "enhancing" his sentence far above the Federal Sentencing Guidelines. The resulting 15-year imprisonment was upheld on appeal by an equally disgusted Court of Appeals. Writing for a three-judge panel, Judge Gilbert Merritt assailed Montgomery's character, pointing out that after a one-month marriage had gone sour, he had sued for custody of his ex-wife's children from a previous relationship.[30]

In another of the biggest scams to come to light, five members of the same extended Alabama family were found guilty of or pled guilty to conspiring to file some 50 fraudulent claims, totaling $3 million, in the names of 37 people. The family members received prison sentences ranging up to 13 years.[31]

The family recruited individuals to file claims stating that they had lost wages working for a company that never actually employed them. The ringleaders assisted the fake claimants in filing the claims and setting up bank accounts where the funds would be deposited. The Gulf Coast Claims Facility paid nearly $2 million of the claims. Fifteen additional individuals who participated in the scheme were also indicted.[32]

The operator of a nail shop franchise in a small-town Walmart in Mississippi pled guilty to submitting a claim in which he pretended to be a fisherman. Although he reportedly received only $18,000 of the $281,000 of the Gulf Coast Claims Facility award, the higher figure was used in calculating his sentence under the Federal Sentencing Guidelines. A Vietnamese immigrant and father of five children, he received a three-year prison sentence, prompting sympathetic coverage under the headline "Feds Hit Hard Against Those Making False BP Claims, Maybe Too Hard" in the business magazine *Forbes*.[33]

BP put itself in the position of serving as a clearinghouse for spotting fraudulent claims and keeping track of the governments' efforts. On its "State of the Gulf" website it offered up a "Fraud Hotline" and kept a "Fraud Tally for Gulf Oil Spill Claims."

Did BP live up to Tony Hayward's promise to pay all legitimate claims? Arguably, BP paid all legitimate *documented* claims that were processed. And once the company agreed to lift the causation requirement for certain claims, it grudgingly paid some of them even if they were not necessarily related to the disaster. No doubt, BP also paid some illegitimate claims that were falsely documented but slipped through the screening process.

As far as the process itself was concerned, there were more hiccups than there should have been. BP overdramatized these flaws, especially given the improvements in screening and the rigor with which fraudulent claimants were pursued by the federal government.

Although the private party claims process yielded considerable controversy, there seems to be little dispute on one point. As the process was winding down in mid-2016, it had cost BP a small fortune—more than $13 billion.

6

THE LEGAL BATTLES
WITH THE FEDS

The BP disaster led to a dozen significant legal cases, criminal and civil, involving the federal government. The first began as early as June 2010, and the last was not resolved until April 2016. Some of the cases were settled, while others were hard fought. The results were mixed.

The Justice Department brought criminal cases against BP and four of its employees; Transocean (the owner/operator of the rig); and Halliburton (the cement contractor) and one of its employees. The federal government also brought civil environmental enforcement actions against BP and Transocean, as well as civil securities law cases against BP and one of its employees. The government defended challenges to the Interior Department's moratorium banning new deepwater drilling in the Gulf and the Environmental Protection Agency's ban of BP from new federal contracts.

In the "win column," the government obtained guilty pleas in all the criminal cases against the corporate defendants, including BP's plea of guilty to manslaughter charges. The government fared less well in the criminal cases against the individuals: no felony convictions or pleas. The civil environmental enforcement action settled after trial for a record $20 billion. The moratorium and contract ban cases settled with no clear winners.

This chapter reviews the outcome of the federal cases, with the exception of the civil enforcement trial against BP, which is discussed at length in the following chapter.

THE CRIMINAL CASES

The government's criminal case against BP was historic in two respects. Rarely is a corporation charged with a homicide, and the charges against BP resulted in a record $4 billion in fines and penalties.

BP pled guilty to seaman's manslaughter, obstruction of Congress, and environmental crimes.[1] The manslaughter charges resulted from the eleven deaths on the *Deepwater Horizon* rig. The obstruction of Congress charges arose from BP's misrepresentation of the oil flow rate after the accident. The environmental crimes related to the negligent discharge of oil and the death of migratory birds.

As is required for any criminal defendant, BP appeared in court in January 2013 to formally enter its guilty plea. In accepting the $4 billion plea deal, Chief Judge Sarah Vance, of the U.S. District Court in New Orleans, came down hard on the company:

> The explosion on the *Deepwater Horizon* rig would never have occurred if BP's employees had properly supervised the negative pressure testing of the well, had not ignored multiple indications that the drill pipe was not secure, had not failed to respond to obvious signs of pressure on the drill pipe, had not failed to contact onshore engineers to alert them of problems, and had not negligently deemed the negative pressure test a success. Their negligence in failing to control the well caused the blowout to occur, which caused the explosion which killed 11 men.[2]

It is not uncommon for corporations to be charged with crimes in the United States, including environmental crimes. But it is relatively rare for a company to be charged with a homicide such as manslaughter. Only in the last few decades have such cases been brought in the United States.[3] Similarly, British law was only recently amended to allow not only corporations but also government departments to be prosecuted criminally.[4] The change was said to be inspired by a tragic train crash in 1999 near Paddington Station in London, in which 31 people died and more than 500 were injured.

One of the first corporate homicide cases brought in the United States involved the Ford Pinto. The state of Indiana indicted Ford Motor Company in 1978 for reckless homicide in the deaths of three high school girls who were in a Pinto that burst into flames after being rear-ended on a highway. The indictment charged that Ford had long known that the 1973 Pinto had an unsafe fuel tank and was reckless in delaying the recall of the vehicle. Ford was acquitted in the legally complicated case.[5]

In 1985, the state of Illinois won an involuntary manslaughter conviction against a company after one of its employees died of cyanide poisoning. The business extracted silver from used x-ray and photographic film using a solution that included cyanide.[6]

What purpose is served by pursuing a corporation criminally instead of civilly when the primary sanction to be imposed in either case is a monetary penalty? The company itself cannot be sent to prison, and its directors, officers, and employees cannot be punished for the company's own criminal acts. Instead, individuals must be separately indicted.

Reasonable minds differ on the question, with some legal scholars taking the view that the criminal justice process is wasted on corporations when civil sanctions are available.[7] Although the concept of double jeopardy does not bar the government from seeking both criminal and civil penalties for the same transgression, arguably there is some overkill in its doing so.

In BP's case, however, there was very little overlap between the criminal offenses and the civil violations. Of the criminal charges brought against BP, only the negligent discharge count also constitutes a civil violation under the Clean Water Act. Moreover, when a company is responsible for such a huge calamity as the BP disaster, arguably it should be subject to both criminal and civil enforcement actions.

Like any defendant with potential criminal and civil liability, BP had an interest in trying to settle both simultaneously in what is known as a global settlement. BP also had an interest in settling the civil actions brought against it by the five Gulf states and consolidated with the federal case. At least one of three major potential stumbling blocks, however, proved insurmountable in attaining a global settlement: (1) disagreement over whether the settlement could be reopened in the event of newly discovered natural resource damage; (2) ambiguity about the amount of the statutory maximum penalty that applied; and (3) the compensation demands of the state of Louisiana.

A re-opener provision, common in environmental cases, allows the government to seek additional relief even after settlement in the event that unknown damage caused by the violation is subsequently discovered. Considering that by reliable estimates as much as 50 percent of the oil discharged into the Gulf remained there when the well was finally sealed, it is hard to fault the government for insisting on such a routine condition of settlement. But, from BP's perspective, it wanted permanent peace and was willing to pay a premium for it.

The amount of BP's civil penalty exposure was based on the extent of its culpability and the number of barrels of oil discharged. As mentioned in chapter one, the maximum civil penalty for *negligence* under the Clean Water Act is $1,100 per barrel. The maximum civil penalty for *gross negligence* is nearly four times higher at $4,300 per barrel. The difference in the legal standard can be subject to interpretation. Simply stated, negligence is carelessness, whereas gross negligence is more extreme carelessness.[8]

The government maintained that BP was grossly negligent in discharging roughly 4 million barrels, whereas BP maintained that it was simply negligent in discharging under 2.5 million barrels. The difference in the statutory maximum penalty between the two positions was some $15.5 billion. This would have been a huge range to overcome in early negotiations considering the statutory maximum often serves as a starting point in discussions.

Louisiana, the Gulf state most heavily impacted by the disaster, took a tough stance on compensation at the negotiating table. One former high-level federal government official who was present attributed the failure to reach a settlement to Louisiana's position. "Louisiana got greedy," she said.[9]

The two other major companies involved in the blowout, Transocean and Halliburton, also faced criminal charges. Transocean pled guilty to the negligent discharge of oil. It was sentenced to a $400 million fine and probation.[10] Halliburton pled guilty to destruction of evidence with respect to the simulations it had run *after* the accident to determine the efficacy of its cement job. It was sentenced to a $200,000 fine (the statutory maximum) and probation, and also made a $55 million "contribution" to the National Fish and Wildlife Foundation.[11]

Individual Prosecutions

In addition to prosecuting BP as a corporation, the federal government indicted four BP employees individually for felonies related to the accident or its alleged cover-up. Some of the cases took until 2016 to be resolved. Two of the employees were acquitted, and the other two pled guilty to reduced misdemeanor charges.

Vice President David Rainey was initially charged with obstruction of Congress (in the investigation by Representative Edward Markey's subcommittee) and making false statements to federal investigators. BP's top two supervisors aboard the *Deepwater Horizon*, Donald Vidrine and Robert Kaluza, were initially charged with manslaughter and an environmental crime. A fourth BP employee, Kurt Mix, an engineer, was initially charged with obstruction of justice.

Mix was the first BP employee to be charged and stand trial for acts relating to the disaster. The government accused him of being involved in BP's misrepresentation of the oil flow rate by deleting text messages of his calculations of a far higher rate than the government and BP were publicly announcing.[12] He was initially convicted of obstruction of justice, a felony, in a trial that was not only hard-fought, but also was riddled with irregularities.[13] Shortly after his conviction, Mix's lawyers tried unsuccessfully to get the judge to recuse himself on grounds that he had personally filed a claim against BP.[14] They did, however, persuade the judge to throw out the conviction as a result of juror misconduct: the jury forewoman had shared with her fellow jurors a comment overheard in the courthouse elevator that influenced her to vote guilty.[15]

The Fifth Circuit Court of Appeals affirmed the district court's decision in June 2015.[16] Instead of retrying Mix on the felony obstruction of justice charge, in November 2015 the government agreed that he could plead guilty to a single misdemeanor count of computer fraud and abuse.[17] He was sentenced to six months' probation. After the plea deal, Mix wrote a *Wall Street Journal* op-ed claiming he had been made a scapegoat.[18]

Rainey, the highest BP official charged, was BP's vice president of exploration for the Gulf of Mexico. After the explosion, Rainey became the deputy incident commander of the BP–Coast Guard "Unified Command" overseeing the response. In that top role, he gave a statement to the FBI and appeared before Congress. He was accused of understating the oil flow rate in the month following the accident.[19]

The government had considerable difficulty making the congressional obstruction charge stick. The trial judge dismissed it long before trial on technical grounds, but the Fifth Circuit Court of Appeals reinstated it on the government's appeal.[20] The trial judge dismissed the charge again at trial, after Rainey's lawyers subpoenaed members of Congress and their staffs, and Congress objected. As a result, only the false statement count went to trial. After very short deliberations, the jury acquitted Rainey. The judge made a point of openly agreeing with the outcome.

Vidrine and Kaluza were the only BP officials charged with crimes directly related to the accident itself. The two were BP's well site leaders, the highest-ranking employees assigned to the *Deepwater Horizon*. They were indicted individually for two types of felony manslaughter in the deaths of the 11 men on the rig. The indictment linked the deaths to the supervisors' mishandling of the results of the negative pressure test. They were also charged with a single count of negligently discharging oil, a misdemeanor relating to the oil discharge.[21]

The district court threw out one set of manslaughter charges and the government eventually dropped the other prior to trial, leaving just the negligence charges. In December 2015 Vidrine pled guilty to the charge. As part of his plea agreement, he was sentenced in April 2016 to 10 months' probation. In February 2016 Kaluza was quickly acquitted of the same charge by a jury.

Why such meager results in the government's criminal cases against the employees? Each case was different, but in each the defendant was presumed innocent and the standard of proof was the same: beyond a reasonable doubt. That proof must include evidence that the defendant acted with the necessary *mens rea* (state of mind). The burden of proof falls squarely on the prosecution, a factor it theoretically takes into account in weighing charges. At trial, the judge instructs the jury to take the burden of proof into account in its deliberations.

Conjecture only goes so far, but some of the cases were hard to prove, if not flawed from the outset, and prosecutors likely debated what charges (if any) to bring. In the Rainey case, the government alleged that Rainey's "best guess" of a 5,000 barrels/day flow rate disregarded advice from a BP engineer questioning the accuracy of the estimate.[22] But, the government was publicly reporting the same rate at the time. The jury and judge that heard the case might well have been influenced by the apparent double standard.

In the Mix case, the government, which recovered many of the deleted text messages, ultimately must have concluded that a misdemeanor charge fit the crime. In the Kaluza and Vidrine cases, the government dropped the involuntary manslaughter counts after a review of testimony in the civil trial led it to determine that it could not meet the legal standard for the charge ("wanton or reckless disregard for human life").[23]

These modest results arguably could be read to suggest that the government pushed too hard initially against the individual defendants. Considering how strictly it has enforced the Clean Water Act in the past, however, the government could have conceivably prosecuted higher-ranking BP personnel, if only for a misdemeanor.

Negligence itself is rarely a crime under U.S. law, but it is a misdemeanor under the Clean Water Act when the act or omission results in pollutants entering water bodies. In one such case, *United States v. Hanousek*, the government successfully prosecuted a mid-level manager who was actually at home asleep at the time of the accident. The defendant was in charge of a rock-quarrying project in Alaska. A forklift driver accidently ruptured a pipeline as he tried to move a rock away from it. Approximately 1,000 to 5,000 gallons of oil flowed from the pipeline into a nearby river. There was no evidence linking the mid-level manager to the accident, other than his general contractual responsibility for the project. Nonetheless, a jury convicted him of negligence, and he was sentenced to six months in prison, six months in a halfway house, and six months of supervised release. The Court of Appeals for the Ninth Circuit affirmed the conviction, and the Supreme Court declined to hear the case.[24]

The Hanousek case, while not binding in Louisiana, is still an important precedent that stands for the principle that managers bear criminal liability for ordinary negligence. Justice Department lawyers likely considered the case carefully as they debated whether to indict BP officials above the level of the two relatively junior ones on the rig itself.

So, why did the prosecutors stop with Vidrine and Kaluza? What about the more senior managers in Houston? Or even London? Perhaps the prosecutorial team thought that the prosecutors in Hanousek had overreached and did not want to do so in such a high-profile case as BP. Prosecutorial decisions also were made years before the November 2015 unveiling of new Justice Department guidelines making the prosecution of corporate officials (in addition to corporations) a higher priority.

Given that Vidrine was present at the time of the accident, it may seem hard to reconcile his light sentence with the relatively harsh sentence for Hanousek, who was home asleep. Sentencing disparities are not uncommon, however, especially between defendants who plead guilty and those that exercise their right to trial. It is also possible that Vidrine got a lighter sentence for agreeing, as he did in his plea agreement, to cooperate with prosecutors and to testify in any trial. He fulfilled the agreement in serving as a witness in Kaulza's trial.

Finally, the Halliburton manager who ordered the concrete job simulations and their subsequent destruction pled guilty to destruction of evidence. He was sentenced to probation and community service.[25]

THE SEC CASES

When the criminal case was settled, so was a civil case brought against BP by the Securities and Exchange Commission (SEC). The SEC sued BP for intentionally misleading investors by repeatedly underreporting the oil flow rate in reports it submitted to the agency. The case was settled in 2012 for $525 million—then the third-largest penalty in SEC history. Its top enforcement official released the following statement at the time:

> The oil spill was catastrophic for the environment, but by hiding its severity BP also harmed another constituency—its own shareholders and the investing public who are entitled to transparency, accuracy, and completeness of company information, particularly in times of crisis. Good corporate citizenship and responsible crisis management means that a company can't hide critical information simply because it fears the backlash.[26]

In a separate action relating to the understatement of the flow rate in the early days following the blowout, the SEC also sued a senior BP employee for insider trading. The employee, Keith Seilhan, an incident commander and on-scene coordinator, directed BP's oil skimming operations and coordinated its oil collection and cleanup operations during the cleanup process. He was accused of another form of skimming—of insider information.

The SEC alleged that during the same period that Seilhan became aware through confidential information that the amount of the oil being discharged was actually greater than the public was being told, he unloaded his entire family stock portfolio in BP—worth $1 million. By selling when he did in late April, Seilhan allegedly avoided some of the subsequent drop in BP's stock price, saving over $100,000 in the process. He agreed to pay the money back and to pay an equal amount in penalties.[27]

The EPA Debarment Case

BP's conduct also subjected it to a process known as debarment, which prevents a company from entering into contracts with the federal government. As one environmental law professor wrote, "If ever a company deserved to be debarred, it is BP."[28]

EPA agreed. Shortly after BP's guilty plea agreement in the criminal case, EPA issued a Notice of Suspension to BP and some 20 of its subsidiaries. As a result, until the suspension was lifted, BP could not enter into any further oil leases or contracts to sell oil to the federal government. The basis of the suspension was BP's "seriously improper conduct" and the need "to protect the public interest" and "insure that the Government conducts public business with responsible persons."[29]

A considerable amount of money was at stake. The following summer, as it ramped up its litigation strategy on a number of fronts, BP sued EPA to lift the suspension. BP apparently sought support from the British government, which filed an *amicus curiae* (friend of the court) brief highlighting BP's contributions to the U.S. and U.K. economies and the perceived effect of the ban on both. "Her Majesty's Government believes that restricting BP's work in the U.S. and for the U.S. Government has the potential to negatively impact significant investment activities, employment, and pensions," the brief said.[30] Both the British government and BP briefs also argued that the suspension should be temporary.

It's not clear if either of these briefs or arguments had much sway, but it is clear that the U.S. government did not want to litigate the debarment case. The case was settled in March 2013. EPA agreed to lift the suspension, and BP agreed to hire a "special EPA monitor" to oversee environmental, ethical, corporate governance, process safety, and educational issues.

The Moratorium Case

In the first major court case involving the federal government following the blowout, the government was the defendant, not plaintiff. On June 7, 2010, offshore oil support companies filed suit challenging the Interior Department's May 28, 2010, moratorium on drilling in the Gulf (the May Directive).

The case was assigned to a New Orleans federal judge, Martin L. C. Feldman, who was sympathetic to the companies' arguments about lost business opportunities and wages. He was also concerned about the ethical implications of the government's misstated scientific support for the moratorium (discussed in chapter 4). The case resulted in an ongoing power struggle between the judge and the government. Ultimately, the judge's own ethics were called into question.

Judge Feldman quickly held a hearing on June 21 and the very next day issued a preliminary injunction barring enforcement of the moratorium.[31] The judge found that the Interior Department had not justified such a broad moratorium on the basis of a single accident, no matter how devastating. Conducting the legally required balancing test of the asserted need for the moratorium in the interest of safety against the financial interests of the parties challenging it, and the larger public interest, the judge found:

> The effect on employment, jobs, loss of domestic energy supplies caused by the moratorium as the plaintiffs (and other suppliers, and the rigs themselves) lose business, and the movement of the rigs to other sites around the world will clearly ripple throughout the economy in this region.[32]

The judge also noted his "uneasiness" about the "factually incorrect" summary that had been in the Interior Department's report: the false implication in the executive summary that the National Academy of Engineering experts had endorsed the moratorium. He called it "a factor that might cause some apprehension about the probity of the process that led to the Report."[33]

In the court hearing, the judge wasted no time getting into the issue of the endorsement issue. The hearing transcript reflects that when he appeared before the judge, the Justice Department lawyer initially tried to explain away the false peer review implication but only a few minutes later was more forthright.

Interviewed some three years later about the case, the judge revealed just how much he was affected not only by the inaccuracy in the report but also by what he perceived as the lack of candor about the subject on the part of the government's lawyers:

The credibility of the experts' reports themselves was a big issue. Credibility is always an issue for a district judge. There had been some misrepresentations, not once but several times in my court, to me, about peer review. These were misrepresentations to me in open court by underlings at the Justice Department, in the presence of the personal counsel to the Interior Secretary."[34]

Judge Feldman's injunction of the moratorium did not stop federal officials. Within days, they appealed and declared that they would issue another moratorium to replace the one Feldman had blocked. On July 12, 2010, the Interior Department rescinded the May Directive and issued another very similar moratorium, albeit one that went further in terms of providing reasoning and evidentiary support. The plaintiffs swiftly came back to court, claiming that the government was in contempt of the judge's injunction. The judge agreed and ultimately awarded over half a million dollars in legal fees to the plaintiffs to punish the government.

It is relatively rare for a federal judge to hold the federal government in contempt of court, but if ever there were a federal agency that should be sensitive to such a possibility, it was the Interior Department. Roughly 10 years earlier, a federal district judge in Washington, D.C., held two successive secretaries of the Interior in contempt in a landmark case brought by Native Americans for the mismanagement of a trust fund under Interior's authority (the second contempt citation was ultimately reversed on appeal).[35]

Fortunately for the Interior Department, the contempt finding in the moratorium case was short-lived. The Fifth Circuit Court of Appeals ultimately reversed it. The appellate court found that technically the district judge's order only barred enforcement of the May Directive and therefore did not prevent the government from issuing the subsequent moratorium on July 12.[36]

Meanwhile, different offshore oil industry companies filed a new lawsuit challenging the July 12 moratorium. Since this was a new case, it was originally randomly assigned to a different federal judge, Kurt D. Engelhardt. Under court rules, however, one case that is "related" to another can be reassigned to the judge already hearing the other case. Given that the plaintiffs and moratorium were different, relatedness was not a foregone conclusion and conceivably could have been litigated by the parties. But there was no chance for that: on July 13 Judge Engelhardt transferred the new case to Judge Feldman.[37]

The new plaintiffs tried to get Judge Feldman to hear their case that summer, but he waited until nearly fall to do so. He then ordered additional briefing. On the very day that additional briefs were due, October 12, 2010, Interior withdrew the new moratorium, which was not set to expire until November 30, 2010.[38]

After the second moratorium had been lifted, the plaintiffs went back to Feldman again to complain that Interior was considering applications for new drilling permits at a snail's pace. Again, the judge found for the plaintiffs, holding that the government was taking too long to process the applications and giving it 30 days to do so. Although the government filed a notice of appeal, it ultimately settled the case by agreeing to review the remaining permits within 30 days and withdrew the appeal.[39]

Having faulted the government for its ethics, the judge's own ethics wound up being scrutinized. The judge's financial holdings prompted environmental groups who had intervened in the first moratorium case to move to disqualify him for having a conflict of interest. Their motion came after the judge had already issued the preliminary injunction, and he denied the motion without a hearing.

According to the evidence introduced at the preliminary injunction hearing, at least 30 oil rigs were affected by the moratorium, including some operated by ExxonMobil. Judge Feldman's lengthy financial disclosure report for calendar year 2010 shows that he held stock in ExxonMobil that year, but sold it on the very day that he issued the preliminary injunction.[40]

In a one-line letter dated the day *after* the hearing and addressed to the administrative office of the U.S. Courts, the judge plainly stated that the stock "was sold at the opening of the stock market on *June 22, 2010*, prior to the opening of a Court hearing on the Oil Spill Moratorium case."[41] The official court record, however, documents that the hearing took place on *June 21, 2010*.[42] Thus, if the stock was indeed sold on June 22, that would mean the judge had heard the case while he still owned stock in a company that at least conceivably had a stake in the outcome of the case.

Judge Feldman's 2010 disclosure statement also showed that he owned—but did not sell—stock in Allis Chalmers Corp., another company that conceivably had a stake in the outcome of the case.[43] That year, Allis Chalmers billed itself "as a multifaceted oil field service company . . . [that] provides services and equipment to oil and natural gas exploration and production companies" including offshore in the Gulf of Mexico.[44] Indeed, on June 30, 2010, the company announced that it had "embarked on an aggressive plan at the end of 2009 to certify and recertify its existing inventory of blowout preventers (BOPs) and components" ahead of the Interior Department's issuance of new guidelines that had been recommended in Interior's May 27, 2010, report to the president.[45]

In a later interview, Judge Feldman steadfastly maintained that he had not created the appearance of impropriety, the ethical standard that applies generally to lawyers and judges alike:[46]

In the Moratorium case, I owned no stock that would be affected by my decision. And the only instance in which I did own stock—which I didn't know it at the time, my broker called me 10 o'clock at night, and I ordered him to sell at the opening of the market. I didn't even know how much I had. I didn't know whether I made or lost money. I just happened to remember reading an exhibit that said that of the 19 companies that had rigs out in the Gulf, Exxon Mobil was one, and I owned a small amount of Exxon Mobil. I didn't feel that there was any valid appearance issue because I didn't own any stock in any company that would be affected by the Moratorium. That's the way I felt.[47]

Appearance issues aside, Judge Feldman's moratorium rulings raise important separation of powers issues. The executive branch was justified in imposing the temporary moratorium in the interest of protecting public safety and the environment until more could be learned about the cause of the blowout. The judge's injunction was unwarranted despite the awkward drafting of one portion of the Interior Department's key report.

MIXED RESULTS

The cases involving the federal government that arose from the BP disaster yielded mixed results. On the criminal side, BP's guilty plea to a homicide, rare as it was, can be considered a big win. On the other hand, the government's failure to secure any felony convictions or prison time for any of the individuals it indicted can be considered a defeat.

Should someone have gone to prison for causing the disaster, or for hindering the investigation of it after the fact? Reasonable people—and prosecutors—will differ. On balance, probably not. As discussed in other chapters, the blowout was an unfortunate accident that resulted from many factors, at least some of which were not the direct responsibility of the employees on the scene. There are probably better ways to prevent such accidents from occurring in the future than making any more of an example of those convicted (as will be discussed in chapter 12). Nonetheless, there is an inherent injustice in imprisoning no one for the accident and 75 people for filing false claims.

On the civil side, the Interior and EPA cases can be considered a bit of a draw, and the SEC case was a relatively small expense for BP. The biggest challenge for both sides was the federal government's civil enforcement action that brought the environmental trial of the century. That case is discussed next.

7

THE ENVIRONMENTAL
TRIAL OF THE CENTURY

The federal government's civil enforcement case against BP was filed in December 2010. The trial that began in February 2013 followed a lengthy period of pre-trial preparation ("discovery") customary in major litigation, as well as a one-year delay in the initial trial date. Other related cases were consolidated in the same proceeding. These included cases brought by the five Gulf states, the private parties, and the other companies that had a role in the blowout.

The trial was divided into three phases. Phase One was to assign liability and fault, including the extent of BP's culpability—critical to determining the level of penalties it owed to the government and the types of damages owed to the private parties. Phase Two was to determine how well control had been lost and how much oil had been discharged into the Gulf of Mexico in the aftermath. Phase Three was to set the final penalty amount.

Not infrequently, a case settles "on the courthouse steps," meaning just before trial. The case by the federal and state governments against BP turned out to be one of the relatively rare cases that settle shortly after the trial.

PHASE ONE

The stormy weather could not have been more fit for the occasion on February 24, 2013, the eve of the first phase of the BP civil trial. Placeholders began to line up shortly after midnight at the federal courthouse on Poydras Street, a short walk from New Orleans's French Quarter. Fortunately, they were shielded from the thunderstorms by the overhang adjacent to the rear entrance of the courthouse. Observers who had followed the case since the blowout nearly three years earlier hired the placeholders to ensure admission into the first-come, first-served seating in the small public gallery, since there were no assigned seats (other than for the lawyers). The trial involving the nation's worst sustainability disaster was not to be missed.

Every seat in the courtroom was filled the next morning when the deputy clerk called "All Rise" and the Honorable Carl Barbier took the bench. A special panel of federal judges responsible for multidistrict litigation had appointed Judge Barbier in August 2010 to preside over most of the civil cases related to the BP disaster, noting that he "has had a distinguished career as an attorney and now as a jurist."[1] Barbier had been a plaintiff's personal injury lawyer with considerable maritime experience before being nominated to the federal court by President Bill Clinton in 1998. Although this experience made him ideally suited for the BP case, his background as a plaintiff's lawyer likely did not endear him to the company.

Federal courtrooms tend to be imposing places, and Judge Barbier's courtroom was no exception. It had the typical design, with the judge's bench elevated in the front, his clerk and court reporter sitting in front of him, and the witness stand off to one side of it.

But the "well" of the courtroom—the area in front of the "bar" or "rail" where the action takes place—had been rearranged for this historic trial. The number of counsel's tables—normally one for the plaintiffs and one for the defendants—had been doubled to accommodate all the lawyers. Since this was a "bench" trial—meaning the outcome would be decided by the judge, not a jury—the jury box was also occupied by lawyers. Others could be found in "overflow" courtrooms.

The number of seats assigned to each legal team roughly corresponded to the size of its role at trial. The Plaintiffs Steering Committee, the team leading the representation of the private plaintiffs and the Justice Department team, occupied most of the seats on the plaintiffs' side of the courtroom. The state government teams, which mostly played a secondary role, had limited seating. On the defendants' side, the BP legal team had the most seats. The remainder on the defendants' side went to the legal teams of the other defendants, including Halliburton and Transocean.

As he opened the trial on February 25, Barbier lived up to his reputation as a cordial, even-handed judge. After the lawyers had introduced themselves, the judge gave a succinct introduction of the case for the benefit of the press and the public. The judge also announced that under an order he had issued the transcripts from the previous Friday, along with the trial testimony and exhibits, would be made available to the public through a website. That order contained a bold-lettered prohibition instructing the parties not to release any trial materials before they were released publicly. (The order spared the Justice Department from having to turn them over to the author of this book, who had earlier filed a Freedom of Information Act request for the trial exhibits.)

The judge's even temperament contrasted sharply with the lawyers' passion as opening statements began. Speaking for the private parties, Jim Roy, a personal injury lawyer from Lafayette, Louisiana, pointed a very heavy finger at the defendants. He called the negative pressure test misinterpretation a "gross and extreme departure from the standards of good oilfield practice."[2] He mocked the failure to notice well "kicks" (well pressure problems) that BP's own expert had said "my Dad would have caught."[3]

Roy continued to use telling quotations, some drawn from BP's own officers, some from expert witnesses. One BP vice president had referred to a "we can get away with this attitude."[4] Another BP vice president had called the accident "entirely preventable."[5] The plaintiffs' process safety expert had referred to a "culture of production over protection."[6] Summarizing, Roy said, "the evidence will show that BP management incentivized a culture of cost cutting, profits over safety, and taking high risk with a conscious disregard for dire potential risks."[7]

Next up, an uncharacteristically flamboyant government attorney, Mike Underhill, reinforced BP's focus on the almighty dollar. "BP put profits before people, profits before safety, profits before the environment," he said. "Time is money, and every dollar counts."[8]

Like Roy, Underhill used BP employees' own words to hang the company. Underhill referred in particular to one employee who had predicted "we are going to get a shitty cement job."[9] "Doing it right and redoing it with a new cement job could have cost delays worth millions of dollars to BP," Underhill said.[10] The company wanted to "squeeze extra profit out of every decision . . . at every fork in the road, BP chose time and money over safety."[11]

Underhill attributed the poor decisionmaking to BP's "corporate culture of recklessness that was chartered in board rooms in Houston and London."[12] Concluding his opening statement, Underhill asserted that BP's actions constituted not just gross negligence, but willful misconduct.[13] This assertion set up the case to later argue for a high per barrel penalty.

Picking up on the grim picture the plaintiffs had painted of BP's ineptitude and greed, the Transocean and Halliburton lawyers also pointed the finger at BP. Halliburton's opening line: "Your

Honor, this case is about a loss of well control and BP's attempts to blame Halliburton and others for BP's misconduct."[14]

BP finally got its chance to rebut that picture late that afternoon as Mike Brock, BP's lead trial counsel, had his turn at the lectern. Brock countered with a courtroom presence and in a soft Southern drawl from his native Alabama that played well in the New Orleans courtroom. He downplayed some of the evidence that had been highlighted by the plaintiffs. Mistakes had been made, he said, not just by BP but also by Transocean and Halliburton. But they were not intentional, and they did not rise to the level of gross negligence or recklessness.

"We do our jobs well," Brock told the judge, using the first person plural as lawyers often do to align themselves with their clients. "We hope we are able to get that over to you during the course of this trial. It was a multiparty, multicausal event. There are a number of different reasons that we have this horrific outcome, but the standards are very high."[15]

Brock's strategy was apparent from the start. He wanted to make the case that the blame for the accident should be shared, not fall solely on BP. He had to concede that mistakes had been made since he was starting the trial with a substantial handicap: his client had already pled guilty to criminal negligence. So, it was a given in the civil trial that BP would be held responsible at least for negligent conduct. Brock's job was to convince the judge that BP's conduct was no worse than that.

Preventing a finding of gross negligence was conceivably a $12 billion job. If the court concluded BP was simply negligent, the statutory maximum penalty was roughly $4 billion, based on the government's discharge figures. If BP was found to have been grossly negligent or reckless, that figure rose to roughly $16 billion.

By the second day of the trial, there were no placeholders, although the courtroom was still full. By the third day, even the casual observer could stroll into the courtroom without a wait.

Both the anti-BP protesters and the television cameras they had attracted the first day had long since disappeared. Such a progressive decline in interest is typical of trials, even historic ones.

As the witnesses started to testify that second morning, the trial quickly evolved into a classic "battle of the experts." Bob Bea, the emeritus professor of engineering at University of California–Berkeley, was the first expert for the plaintiffs. While he looked and played the part of an erudite academic, complete with silver hair and mustache, academia was a second career for Bea. During his first career, he worked for Shell for sixteen years, including on an oil rig, and had started a couple of his own engineering firms.

Bea was guided on direct examination by questions from Bob Cunningham, a Plaintiffs Steering Committee lawyer and the lead name partner in a Mobile, Alabama, firm. His firm's profile reflects that he flew more than 500 combat missions as a helicopter pilot during the Vietnam War. Bea testified that the disaster was the result of failures in process safety (the prevention of catastrophic accidents), about which he had consulted for BP over the course of five years a decade earlier.

Bea was subjected to a withering cross-examination by Brock that was so thorough that the judge asked him to "move this along" at one point.[16] Brock scored some points by getting Bea to admit that BP had taken positive steps to improve its process safety, including making a major commitment in the years following Bea's consulting work for the company. Bea countered: "Well, the statement of the talk has to be backed up with effective walk."[17] Brock's effectiveness that day won over the judge who later ruled that BP's process safety management system "may not have been perfect, but the evidence has not shown that it was defective or a cause of the blowout, explosion, and fire."[18]

The next expert witness for the plaintiffs was Alan Huffman, a geophysicist with extensive offshore mining and drilling experience. He gave well-rehearsed testimony elicited by an equally well-prepared Harvard-trained Justice Department lawyer, Daniel Spiro. Huffman testified in no uncertain terms that the well had been drilled in an unsafe and dangerous manner. In particular, he said BP's decision to drill an extra 100 feet after it had already reached 18,260 feet, was "egregious beyond anything I have seen in my career."[19]

Huffman's testimony went into deep detail about how BP had repeatedly violated industry practice safety margins—and federal regulations. Judge Barbier followed closely, even interrupting at times to ensure that he was understanding correctly. When the judge ruled in September 2014, it became clear just how influenced he had been by Huffman's testimony. The judge noted that on the whole he agreed with Huffman's conclusions, and quoted portions of it.[20]

To counter Huffman's damaging testimony, BP put its own drilling expert on the stand, Louisiana State University emeritus professor of petroleum engineering and former dean Adam Bourgoyne. When asked whether he agreed with Huffman's opinions, Bourgoyne testified repeatedly, "I totally disagree with that" and "I don't agree with that at all."[21] Although the judge found Bourgoyne well-qualified and agreed with him to a degree, the judge gave him much less credit than Huffman and ultimately disagreed with Bourgoyne's conclusion that the unsafe drilling could not be linked to the blowout.[22]

Additional experts testified about the cementing of the well and the all-important negative pressure test. As expected, there were differing explanations for why the cement used to seal the well failed, but all of the experts agreed that if there is pressure or flow on a drilling line, then the test reflects failure. In the words of government expert Richard Heenan, this is a "pass-fail" test—"fail, to be exact," in this case.[23]

The plaintiffs also called a number of witnesses, including BP top brass. First up was Lamar McKay, president and chairman of BP America at the time of the blowout. He was questioned by Bob Cunningham. Although he may have been a wartime ace, the lawyer failed to land any shots on McKay. In a high-stakes case like BP, it would not have been atypical for the company's lawyers to have spent the better part of a week preparing an important witness like McKay to testify. That preparation showed.

The plaintiffs also scored few points with the testimony of former CEO Tony Hayward. By then outside the jurisdiction of the court, he was not legally compelled to appear at trial. So the plaintiffs introduced instead a videotape of his earlier deposition testimony. Hayward also had been well prepared. In fact, it took quite some time for him to even admit to some leadership principles that he had personally espoused at a speech he had given at Stanford Business School.

Some of the most riveting testimony related not to the underlying causes of the blowout itself, but to its immediate aftermath on the rig on the night of April 20, 2010. Roughly a month into the trial, the first mate of the *Deepwater Horizon*, David Young (called as a witness by Transocean, his employer), testified firsthand about the chaos on board in the minutes following the explosion.

Young had come across one of his shipmates, Dale Burkeen, whom he took for dead, and later had to be physically restrained from going back to get him out. At first, Young started laying out fire-fighting equipment, but it was soon clear to him that the ship would have to be abandoned. He said he was not on the bridge to hear sometimes expletive-filled conversations between the captain and other crew members about whether to detach the rig from the well or send mayday signals. Young seemed proud that 115 crew members had escaped with their lives.

Other testimony was not nearly as exciting. Like at most trials, especially complex ones, much of the testimony tended to be pretty dry. It is not uncommon for jurors—and even lawyers and judges—to fall asleep. In the sixth week of the lengthy Phase One trial, the testimony apparently did lull one lawyer to sleep.

Early in the trial, Judge Barbier had chided Alan Kanner, a private attorney representing the state of Louisiana, for making statements instead of cross-examining the witness when Kanner thanked Bob Bea for his entirely unrelated work on levees in Louisiana following Hurricane Katrina.[24] On April 11, apparently believing that Kanner was dozing off, the judge interrupted testimony about battery testing to say "Would someone tap Mr. Kanner and wake him up?"[25]

At this point, only three more days remained in the trial, after which the lawyers could catch up on their sleep before submitting their post-trial briefs over the summer. In its brief, BP argued that the blowout "resulted from a series of independent acts and omissions by independent individuals that, when combined, had the effect of overcoming state-of-the-art safety systems."[26] Noting the multiple parties involved in multiple chains of events, BP contended that its conduct alone "does not remotely approach [the] demanding standard" necessary to establish gross negligence or willful misconduct. That, BP said, required "a particular defendant, with a culpable mental state, engaged in extreme and egregious misconduct."[27]

As expected, the federal government had quite a different take. Its brief argued that gross negligence or willful misconduct "can be proven by showing either a single act or omission, an accumulation of acts or omissions, or both."[28] Under the single act or omission theory, BP's bungled negative pressure test alone was enough to constitute gross negligence or willful misconduct. Under what it called the accumulation of acts (or "Swiss cheese") theory, the government presented a bullet list of eight factors that contributed to the blowout, which it said together also constituted gross negligence or willful misconduct.[29]

The post-trial briefing schedule that the judge had set in the summer of 2013 led to speculation that he would rule on the critical negligence issue by Labor Day. He did—by Labor Day of the *following* year. The judge finally issued his Phase One Decision on September 4, 2014.[30]

At 153 pages, the judge's Findings of Fact and Conclusions of Law were articulate and very thorough. The judge ruled that BP's actions constituted "gross negligence" and "willful misconduct" under the Clean Water Act, adopting both the government's single and multiple act arguments.[31] The judge found that the reckless actions of BP employees with respect to the negative pressure test—misinterpretation of the results and failure to order a new test—alone made BP liable for willful misconduct and gross negligence.[32] He further found that BP "committed a series of [eight] negligent acts or omissions that resulted in the discharge of oil, which together amount to gross negligence and willful misconduct under the CWA [Clean Water Act]."[33] Because the judge found that BP's conduct had risen above the level of ordinary negligence, the company was liable to the federal government for up to $4,300 per barrel of oil discharged.

Finally, for the purpose of the private party claims, the judge found that BP had also been "reckless" within the meaning of general maritime law, whereas Transocean and Halliburton had only been "negligent." The judge apportioned 67 percent of the fault to BP, 30 percent to Transocean, and 3 percent to Halliburton.[34] He also found that although the conduct of BP's employees leading up to the accident was egregious enough to warrant punitive damages, under Fifth Circuit precedent, BP was not liable for the conduct of its employees.

PHASE TWO

Phase Two of the trial started on schedule on September 30, 2013. The first week focused on the question of "source control": Did BP's efforts to contain the oil stream after the explosion warrant the award of punitive damages to the private party plaintiffs? The remaining two weeks focused on "quantification" for the purposes of calculating the maximum civil penalty: how much oil was discharged from the well? The federal government argued that

5 million barrels had escaped from the well. BP said it was closer to 3.25 million. Both sides did agree that BP managed to siphon off 810,000 barrels as it flowed from the well before it hit the ocean, so that amount was to be deducted from the total for which BP was liable.

Compared to Phase One, Phase Two was anticlimactic. As the judge later observed, "The evidence pertaining to the Quantification segment was voluminous, dense, highly technical, and conflicting."[35] Indeed, virtually the only excitement was when, at the end of the first week, the judge encouraged the parties to wrap up promptly and get out of town as a hurricane brewed in the Gulf. The hurricane (Karen) didn't materialize.

The judge issued his decision on Phase Two on January 15, 2015. First, the judge found that the private plaintiffs were not entitled to punitive damages since BP's efforts to control the well after it blew out had not been unreasonable, and its planning for such an event had complied with federal regulations and industry practice. Next, after summarizing the complex quantification evidence in just four short pages, the judge found that 4 million barrels had been released from the well. Without further explanation of how he had reached that figure, he simply deducted the 810,000 barrels that had been "collected" at the wellhead and found that 3.19 million barrels had been discharged into the Gulf.[36] At the $4,300 per barrel statutory maximum applicable as a result of the Phase One ruling, this meant that BP was liable for up to $13.7 billion in penalties. It would be up to the judge to decide in Phase Three how much of that amount to penalize BP.

PHASE THREE

Phase Three of the trial began on January 20, 2015, nearly two years after Phase One. Phase Three pitted the government against BP over how much the company should pay in civil penalties. Since the Phase Two ruling just five days earlier on the amount of the discharge affected how the lawyers would argue the penalty, they were occasionally recalibrating their arguments on the run during Phase Three.

Under Clean Water Act case law, the judge has considerable discretion in imposing a civil penalty, so long as he applies the eight factors mandated by the statute:[37]

1. The seriousness of the violations;

2. The degree of culpability involved;

3. The efforts of the violator to minimize or mitigate the effects of the discharge;

4. Any history of prior violations;

5. Any other penalty for the same incident;

6. The economic impact of the penalty on the violator;

7. The economic benefit to the violator, if any, resulting from the violations;

8. Any other matters as justice may require.

The eighth factor is a catchall that allows the judge to take into account other considerations. For example, Judge Barbier could have assessed the importance of imposing a high penalty for deterrent value, as his colleague, Judge Sarah Vance, had done in the criminal case.

Both sides were well-staffed with lawyers in Phase Three, including some new faces on the government's side occupying both counsel's table and the wing seats. In stark contrast to BP's legal team, six of the ten government attorneys were women, including four of the six attorneys seated at the government's table.

In his opening statement, Steve O'Rourke, for the government, went methodically through the eight penalty factors, emphasizing the seriousness of the event and BP's substantial wealth. BP had paid $19 billion in dividends since the disaster, he pointed out. That was considerably more than the $11.7–13.7 billion range he suggested as a penalty.

In his opening statement, Mike Brock for BP did not exactly plead poverty on the company's behalf, but he did assert that the court should not look beyond BP Exploration and Production (BPXP), the BP subsidiary actually named in the government's case, to pay the civil penalty. The lawyer pointed to the steep drop in oil prices in prior months to justify a major reduction in the penalty. And he pointed to what he called the company's expensive response efforts.

As Brock emphasized that BPXP had suffered a vast reduction in its fourth quarter 2014 valuation due largely to the steep decline in oil prices, the judge interrupted to ask: "Is there any reason that any penalties can't be structured to be paid over a number of years?"[38] The judge's unusual interjection could be read to suggest that he might be considering such a substantial penalty that, from the standpoint of financial impact on the company, payments would need to be a structured over time. It was a good sign for the government, and O'Rourke was understandably happy about it.

The opening statements were followed by the government's witnesses, beginning with a Coast Guard admiral who had supervised part of the response. She was generous in the credit she gave BP for its efforts.

Judge Barbier became a little impatient with both sides very early in Phase Three. BP had filed objections to the legitimacy of the testimony of every one of the government's expert witnesses. This tactic led the judge at the opening of the second day to observe that BP had undermined its own credibility in doing so.[39] By the end of the third day of trial, the judge was encouraging the government to work on limiting its direct examination lest the testimony "put the jury [meaning the judge] to sleep."[40]

If that was a hint, the lawyers all got it. Phase Three, originally scheduled to take three weeks, took only two. On February 2, 2015, the trial involving the biggest environmental disaster in American history ended. Complimenting the lawyers for their professionalism and civility, the judge said it had been a pleasure for him to try the cases. Court was then adjourned.

THE DEAL

There were signs that the judge would rule during the summer of 2015, including the abrupt cancellation in mid-June of a routine status conference pending issuance of his decision. Then, before the judge had ruled, the case was settled.

On July 2, 2015, after two years of intermittent trial spanning from February 2013 through February 2015, BP announced that it had reached a mammoth agreement in principle with the federal and state governments. BP's timing might have been geared for the quiet news cycle in advance of the July 4 holiday weekend, or in anticipation of the issuance of its second quarter business report.

The deal, as portrayed by BP, consisted of a $18.7 billion settlement with the federal and state governments. The breakdown was: $7.1 billion in natural resource damages (in addition to $1 billion that already had been committed); $5.5 billion in Clean Water Act civil penalties; and $5.9 billion in state and local economic damages ($4.9 million to the Gulf states and $1 billion to local Gulf entities). An additional $232 million would be paid for unknown natural resource damage. Payments would be spread over a lengthy period of time ranging from 15 to 18 years, with interest on the natural resource and civil penalty payments. BP said the deal was costing it roughly $1.1 billion a year.

Following BP's announcement, the Justice Department touted: "If approved, this will be the largest environmental settlement in the history of the United States, and the largest civil settlement with a single entity ever by the Department of Justice."[41]

Although the parties had agreed on a deal, they had not yet agreed on all the details. This took a few more months. On October 5, 2015, with the written agreement signed and lodged with the court, the Justice Department held a celebratory press conference at its Washington, D.C., main building.

The attorney general, Loretta Lynch, presided, with the secretaries of commerce and agriculture, EPA administrator, the commandant of the Coast Guard, and the deputy secretary of the Interior at her side. The attorney general highlighted not only the price tag of the settlement, but also the launch of one of the largest restoration efforts the world has ever seen. She portrayed the deal as being worth $20.8 billion.[42]

The $2.1 billion discrepancy in valuation between BP and the government was subtle, but not substantive. The government had totaled up every conceivable payment, including the $1 billion already advanced for early restoration and monies that might be due for unknown damage. The terms of the deal had not changed since BP's announcement in July—only the presentation of it.

After a "notice and comment" period during which the public had the opportunity to submit its views on the settlement, on March 22, 2016, the Justice Department filed a motion formally requesting that the judge approve it. The judge promptly did so on April 4, 2016, by putting his signature on the consent decree and entering a one-line judgment.[43]

The environmental trial of the century had concluded with the deal of the century. Both sides had reasons to be pleased. The federal and state governments had leveraged another $20 billion from BP, much of which would go to the Gulf. BP had made amends, and had obtained the closure it needed to move on with its business.

The settlement was indeed record-breaking, in keeping with the scale of the disaster. The $5.5 billion penalty represented roughly 40 percent of the statutory maximum civil penalty of $13.7 billion. Even for an egregious case, the penalty was very substantial given that Clean Water Act penalties rarely even approach the statutory maximum. Still, the penalty was far less than the Justice Department asked the judge to award at trial. If the case had not settled, he might have imposed a higher penalty, especially if he took into account BP's average annual profit in the four years following the blowout: nearly $16.5 billion. The $5.5 billion in civil penalties represents one-third of that average. (By comparison, the jury's

$5 billion punitive damage award in the *Exxon Valdez* fishermen's class action two decades earlier represented one year's profits, although it was repeatedly overturned by the appellate courts.) BP certainly could have afforded more. On the other hand, if the judge had focused solely on the finances of BPXP (and not those of the parent company), he might have imposed a smaller penalty than was negotiated.

An assessment of the natural resource damage component of the settlement is more difficult. Since BP's natural resource damage liability was not part of the case that was tried, a second trial would have been necessary, fraught with even more uncertainty for the parties. Because natural resource damage cases are inherently complex, they are rarely tried. Proving the natural resource damages that result from an event as massive as the BP disaster would have been extraordinarily challenging for the governments, so in principle it made sense to settle that aspect of the case. The state and local governments also appear to have reached a reasonable compromise on their economic losses.

On balance, the settlement was a fair deal for both sides. There is reason to be concerned about it, however, from the taxpayer's perspective. The hidden cost of the settlement (and the disaster as a whole) to the taxpayer will be discussed in the next chapter.

8

THE COST

By mid-2016 BP had estimated its disaster-related costs at $61.6 billion and all but closed its book on the disaster by declaring that it did not expect any further claims to materially affect the company's financial performance. The largest of the many sets of costs were $25 billion in government settlements, $14 billion in reported response costs, and $13.5 billion in private economic and property claims. The costs appear to well exceed any ever paid by a company for a manmade disaster.

It is fairly straightforward to measure in broad terms how much the disaster cost BP since that figure is reasonably tangible, and is one that BP tracks and reports. Theoretically, those costs are borne by BP's owners—its shareholders. However, because most business expenses are tax-deductible, the American taxpayer will share the burden, essentially subsidizing a substantial portion of them. The actual net effect of such deductions will vary with BP's actual tax rate for any given year, but given BP's tax rates in prior years, the taxpayer could be subsidizing roughly one-third of BP's deductible business expenses. Thus, if, as expected, about $52 billion of BP's costs are considered deductible, that means that the taxpayer essentially could be absorbing approximately $17 billion of them.[1]

It is much harder to measure the cost of the disaster to society as a whole. While lost livelihoods such as business revenue and wages can be quantified with reasonable accuracy, lost lives and the effect of illness cannot. Nor can the value of lost natural resources. The disaster also produced many other hidden societal costs. The total costs to society may never be known.

BP'S COSTS

BP incurred massive costs in the months immediately following the blowout as it worked to plug the well and contain and clean up the oil coming from it. The cleanup and containment operation required vast amounts of resources, including personnel, equipment, operational building space, aircraft, boats, and much more. BP reported that its total response costs exceeded $14 billion.

Without a comprehensive breakdown by BP of how the $14 billion was spent, it is difficult to assess exactly what the figure covered. For example, does it include costs incurred in launching BP's initial post-accident public relations campaign, with all its expensive television commercials?

What is known is that in addition to paying its own expenses, under the Oil Pollution Act, BP also needed to reimburse the federal, state, and local governments for the costs they incurred in responding to the disaster. Roughly $1.5 billion went for this purpose.

Anticipating a major financial liability as a result of the disaster, BP took a charge against its earnings of nearly $41 billion in 2010. The charge put the company into negative profits for that year, but BP showed healthy profits averaging over $20 billion a year in the following three years, before profits declined steeply as oil prices dropped beginning in 2014. In 2015, the year of the state and federal civil settlement, BP took another major charge against earnings of nearly $12 billion, putting its profits again in the loss column. At the end of the second quarter of 2016, BP announced another such charge of $5.2 billion, bringing

BP's estimated costs as a result of the disaster above the $61 billion mark. At that point, the company said that it did not expect future disaster costs to have a material impact on its financial performance. In other words, it believed it was finished with the major expenses.[2]

BP's anticipated expenses were reflected in the company's loss in share value. BP's stock plummeted immediately after the blowout. Hovering around $60 on the New York Stock Exchange when the well first exploded, the share price steadily dove as the oil streamed out, hitting a low of $27 in late June 2010. Thus, at the lowest point in the stock drop during the 87-day period before the oil well was capped, the company's share price had declined by more than 50 percent.

A company's value is often measured in "market capitalization," a function of the number of shares outstanding multiplied by the share price. When the well was plugged in mid-July 2010, the stock had recovered to around the $39 level. The loss in the company's market capitalization at this point was measured at more than $67.5 billion. Although the stock subsequently rose, using sophisticated modeling, University of Louisiana economists found that a year after the accident, the stock was still suffering from what they called a cumulative abnormal return of nearly negative one-third (of the price it otherwise would have been).[3]

Who were the major shareholders at the time that took the blow? "BP may be British, but Americans own half the stock," according to a June 2010 story in the *New York Times*.[4] They included American investment firms and pension plans. Hardest hit was BlackRock, the largest investment firm in the world, based in New York City. It reportedly held nearly 6 percent of the BP stock outstanding at the time of the explosion.

Whether they were American or another nationality, BP shareholders did not take the losses lightly. Institutional shareholders filed a barrage of lawsuits based on a variety of legal theories. They met with mixed success.

In one of the most significant shareholder cases, New York State and Ohio employee pension funds holding BP stock sued the company for securities fraud. A federal judge granted class action status but limited the class members to purchasers of BP stock *after* the blowout. The theory of the suit was that these shareholders suffered losses because BP misled investors about the oil flow rate following the blowout. Under the district court's decision, affirmed by the Fifth Circuit Court of Appeals, purchasers of BP stock through U.S. exchanges for roughly the 13 months following the explosion could join the lawsuit.[5] The case settled in June 2016 for $175 million.[6]

Another shareholder securities fraud suit was brought in Texas by British and other foreign investors who bought BP stock in Great Britain. They, too, allege that BP stock went down further than expected after the explosion because BP misled them about the severity of the accident.

BP employees in North America whose pension fund invests entirely in BP shares brought a classic "shareholder derivative suit" in which investors claim to be suing on behalf of the corporation itself. They asserted that BP officers and directors breached their fiduciary duties to BP. Although the trial court initially dismissed the case, the Fifth Circuit Court of Appeals reversed that decision.

Another shareholder derivative case was brought under British law in Houston. This one was dismissed by a federal district judge who ruled that it should be heard in Great Britain. The Fifth Circuit Court of Appeals affirmed that decision.

The shareholder litigation was consolidated into multidistrict litigation overseen by a federal district judge in Houston, home of BP's U.S. headquarters. Some of it drags on. If the remaining cases are successful, the cost of the disaster to BP conceivably could grow. And even if the cases do not ultimately succeed, BP's legal fees will continue to mount.

On the other hand, BP is not at further risk of incurring costs related to lost business claims precipitated by the 2010 moratorium, such as those brought by offshore oil services companies. In March 2016 Judge Barbier entered an order in a lawsuit brought by these companies holding that BP was not liable for their losses. The judge found that the moratorium, although precipitated by the Macondo well blowout, was designed to address the potential for blowouts from other wells. Therefore, the resulting losses were not BP's to bear. This was a big win for BP. Had it been held liable for the moratorium costs, its additional costs could have been in the billions of dollars. Two oil services firms reportedly repeatedly blamed the moratorium for their bankruptcies, and others alleged nine-figure losses.[7]

TAXPAYER COSTS

The American taxpayer likely will wind up financing a substantial percentage of BP's disaster-related expenses. As noted, with limited exceptions, business expenses generally are tax deductible under U.S. law. There is every reason to believe that BP will fully avail itself of that opportunity with respect to the roughly $52 billion of its expenses that appear to be deductible.

The Settlement Payments

The reason not all of BP's expenses are deductible is the exception for the payment of civil and criminal penalties. By law, these payments cannot be deducted, and both the criminal guilty plea agreement and civil settlement agreement contain language to that effect. The criminal settlement was just over $4 billion, and the penalty component of the $20 billion civil settlement was $5.5 billion. Thus, $9.5 billion cannot be deducted.

The deductible portion of the civil settlement, then, is $14.5 billion. Allowing BP to deduct that sum did not sit well with some members of Congress, among others. Senator Bill Nelson of Florida wrote Attorney General Loretta Lynch shortly after the deal was first announced asking her to ensure that the settlement agreement specifically provide that its payments will not be deductible. During the subsequent formal notice and comment period, more than 50 members of the House of Representatives wrote the attorney general asking her to insert language barring BP from deducting any portion of the $20 billion deal. Nearly 29,000 (99 percent) of the comments that the Justice Department received about the settlement also objected to the deductability of 75 percent of the settlement payments.

The Justice Department essentially punted on the deductibility issue when it formally requested that the court enter the decree: "If and when BP entities present their position on the taxable status of any payment under the Consent Decree (other than penalty payments), the relevant tax authority will make a determination in accordance with applicable law. This is the consistent process taken in all similar consent decrees."[8]

BP likely factored in the tax implications when making the deal. Had the government pushed to include a clause prohibiting the company from entirely writing off the deal, BP likely would have pushed for a smaller settlement. Still, the practice of allowing companies to receive taxpayer subsidies for the consequences of their pollution needs to be addressed, whether by the Justice Department, the IRS, or Congress.

The Payment Plan

Another potentially troubling issue from the taxpayer perspective is the favorable payment plan that BP received under the settlement agreement.

The federal civil penalties and natural resource damage payments will be made over a period of fifteen years, with interest. More than $11.5 billion of the funds will go to Gulf restoration (damage repair). That includes the additional $7 billion specifically allotted for that purpose, as well as 80 percent of the civil penalties.

Staging payments over a period of years makes sense since the restoration money will be used for projects that will take time. Whether BP should have received a highly favorable interest rate for the payments is less clear. Neither side disclosed the actual interest rate for the payment plan in its press releases or attachments. But based on a footnote in BP's press release describing the benchmarks for establishing the rate, it was calculated by an economist to be .82 percent (that is, less than 1 percent), compounded annually.[9]

The low interest rate was quite a coup for BP. Other benchmarks that could have been used would have resulted in a much higher rate. The prime rate (the rate that banks charge their best customers, usually big corporations) at the time was 3.25 percent, roughly the same as the average rate for a 15-year mortgage.

The payment plan and favorable rate that BP received were exceptions to the federal government's usual practice. While the government and BP were finalizing the settlement in principle, EPA was finalizing its formal guidance on the settlement of administrative enforcement actions (cases handled at the agency, not at the judicial level).[10] In instructions to its litigation teams, the agency said that payment plans preferably should not exceed three years and should use either the prime rate or the IRS underpayment rate (5 percent at the time for large corporations). EPA's official Clean Water Act Settlement Policy also says installment plans "should generally should not extend beyond three years," with "appropriate interest accruing on the delayed payments."[11]

BP received a very good financing deal, again, arguably at taxpayer expense. The difference between the market interest rates and the rate BP received was about 2.5 percent. Applying that difference to the delayed payments in the civil settlement deal amounts to a very substantial savings—well in excess of $1 billion.[12]

In its brief in support of its motion for the consent decree, the Justice Department readily acknowledged that the payment plan represented a departure from the norm. The government argued that the exception was justified by the "specifics of this case" and the "absolute magnitude of the sums paid."[13] The government also hinted that it had made a tradeoff in accepting the payment plan in exchange for a higher penalty: "basic negotiation practice also informs that a defendant will pay a higher dollar amount if allowed to do so over a longer period of time."[14] The brief was silent on the interest rate.

The Cap on Unknown Future Damage

Finally, under the settlement, the taxpayer bears some of the risk of having to fund restoration of natural resource damage that was unknown at the time of the settlement. The usual government practice in these types of cases is to require settlements to contain what is known as a "reopener clause." As the term suggests, such a provision essentially allows the deal to be reopened and the defendant to be liable for additional costs in the event that unknown natural resource damage is later discovered. From a legal standpoint, a re-opener makes sense since if a natural resource case did go to trial, the defendant would not be immune from liability for later discovered damage.

Again, the BP consent decree strays from the norm, potentially at the taxpayer's expense. Rather than being exposed to unlimited additional liability for unknown natural resource damage, BP's liability is capped at $700 million above the $8 billion in required natural resource damage payments.

Although unlimited reopener clauses are customary, there is precedent for caps in major cases. In the settlement with the government in the *Exxon Valdez* case, for example, the cap was $100 million, on top of payments of $900 million in natural resource damages. Thus, the percentage of the cap relative to the natural resource damage payments in the BP case is roughly consistent with the percentage in *Exxon Valdez*.

There is no precedent, however, for gauging the long-term effects on the ecosystem of such a substantial deepwater ocean discharge as occurred in the BP case. Thus, the risk of incurring undiscovered natural resource damage above the $700 million level is potentially high. Juxtaposed against this risk, the Justice Department contended that there was litigation risk in trying to prove that any later discovered natural resource damage was attributable to BP's oil.

Given reports that the re-opener had been a bone of contention in pre-trial negotiations, the cap might well have been an essential part of clinching the deal with BP—and therefore a concession that the government was willing to make. As a result, the taxpayers run the risk of bearing the cost of any undiscovered natural damage exceeding the $700 million cap. Only time will tell whether the government gave away too much or made a good deal. Considering all the unknowns associated with such an unprecedented discharge, however, deviating from standard practice in the BP case is somewhat difficult to justify.

SOCIETAL COSTS

The costs of the disaster to society at large were both profound and difficult to quantify. These societal costs include human life and health at one end of the spectrum and lost recreational opportunities at the other. They also include the diminished value of the "natural capital" provided by damaged natural resources.

Just as life is said to be priceless, how does one measure in dollar terms all of the effects that the disaster brought to the daily lives of the Gulf's inhabitants and visitors? How can a value be placed on the disruption of one's daily life, whether that means increased anxiety, a lost beach vacation, boating trip, fishing trip, or sporting event?[15] Consider, also, the lost aesthetic value of a once pristine beach turned into an oily mess, if only temporarily.

The nonmarket value of natural resources is particularly difficult to quantify. One study estimated that the annual natural capital from the Mississippi Delta ecosystems should be valued in the $12–$47 billion range.[16] The study assessed the annual benefit of goods and services including hurricane and flood protection, water supply, water quality, recreation, and fisheries.

A substantial portion of the Mississippi Delta ecosystem, which comprises an area of some 520,000 acres at the mouth of the Mississippi River off the Louisiana coast, was harmed by BP's oil. Even if BP is said to be responsible for only a limited reduction in the ecosystem's annual natural capital valuation, the dollar figure attributable to the disaster could add up to tens of billions of dollars over the 15-year restoration payment period.[17] The potential annual cost of lost hurricane protection alone attributable to wetlands depletion could be in the billions of dollars—and could soar in the event of a major hurricane. Such societal costs—and many others—might be very difficult to calculate with a significant degree of certainty, but are still very real.

THE MOST EXPENSIVE CORPORATE MANMADE DISASTER

The nation's worst offshore oil discharge has resulted in what appears to be the world's most expensive manmade corporate disaster. At $61.6 billion, BP's estimate of its total costs broke all known records.

Significantly, the taxpayer bears all of the risk of any unknown natural resource damage costs that exceed the $700 million cap. Depending on those potential costs—as well as how other societal costs are valued—all told the cost of disaster might wind up growing substantially.

No matter how one values the costs of the BP disaster, they were enormous. Enormous for the company, its shareholders, the American taxpayer, and society as a whole. BP may have all but closed its books on the disaster, but the taxpayer and society may be left holding the bag.

9

THE SPIN GAMES

As oil flowed uncontrollably into the Gulf, back in Washington the Obama administration struggled to control its political image. The White House was on the defensive for three reasons: (1) only three weeks before the blowout, the president had announced plans for the expansion of offshore drilling; (2) the federal government had failed in its job of regulating offshore drilling; and (3) the federal government was poorly equipped to manage the disaster.

The last thing the White House wanted was for the disaster to be "Obama's Katrina," as some critics were calling it in a reference to President George W. Bush's failure to respond effectively to the 2005 hurricane. The Obama White House wanted desperately to avoid the dire missteps of the Bush team during Katrina. But despite being acutely aware of how badly it was being portrayed in the media, the administration faltered repeatedly in trying to project a positive image.

The administration sometimes seemed to be at odds with itself in terms of its spin strategy. On the one hand, the administration wanted to show how seriously the president was treating the disaster, such as by elevating it to the status of a national security threat. On the other hand, the administration seemed to be downplaying its seriousness by offering up unduly optimistic reports of both the amounts of oil being released and captured.

Meanwhile, BP labored to control its corporate image. It, too, was on the defensive because of past and current events: (1) it had a very poor environmental and safety record; (2) as the well owner, it was responsible for the blowout; and (3) it was ill-prepared to respond to the disaster. The last thing BP wanted was for the disaster to become CEO Tony Hayward's "Texas City"—a reference to the explosion of a BP refinery in 2005 that caused the deaths of 15 workers and damaged the reputation of the previous CEO, Lord John Browne. BP's strategy also walked the tightrope between acknowledging the seriousness of the problem and downplaying it by understating the flow rate.

Played out at the highest levels, the spin games of both the administration and BP only made them look worse at times.

THE DECISION TO EXPAND OFFSHORE OIL DRILLING

On March 31, 2010, President Obama gave a major speech in which he announced a significant expansion of offshore drilling. The speech took place at Andrews Air Force Base outside of Washington, D.C. Later renamed Joint Base Andrews to reflect the presence of a Naval Air Facility, the base functions primarily as the airport for Air Force One and other "MILAIR" flights—military aircraft used for high-ranking civilian and military dignitaries.

The venue was carefully chosen, as were the president's words, to link the expansion of offshore drilling to national security. Obama said that his administration would be lifting bans on offshore drilling along the East Coast from Delaware to the central coast of Florida, as well as in the eastern part of the Gulf of Mexico and portions of Alaskan waters. He conceded at the time that many would disagree with his decision.

The March 31 announcement caught many environmentalists off guard since the president had campaigned on a "clean energy" platform and then made climate change legislation a high priority. The announcement also reportedly caught some influential senators by surprise, including Senators John Kerry of Massachusetts, Joseph Lieberman of Connecticut, and Lindsey Graham of South Carolina—the three key members involved in negotiations over the Senate version of the climate bill. Senator Bob Menendez, a New Jersey Democrat on the Energy and Natural Resources Committee, reportedly left an angry message that night on the cell phone of Interior Secretary Ken Salazar.[1] The expansion of offshore drilling had been a bargaining chip during the ongoing climate bill negotiations. Negotiators feared they could no longer get anything in return for the concession.

High-level environmental officials in the executive branch later testified that they also had not been consulted about the expansion decision. They included the administrator of the National Oceanic and Atmospheric Administration (NOAA), Jane Lubchenco, and the chair of the Council on Environmental Quality (CEQ), Nancy Sutley.[2]

Why would key players not have been consulted, and on whose advice did the president instead rely? In his speech, Obama said that the decision was made after careful deliberation by Carol Browner, his chief energy and climate adviser, and Salazar, whom he described at the time as "one of the finest secretaries of Interior we've ever had."[3] Both Salazar and Browner were present at the announcement, as were Secretary of Energy Steven Chu and CEQ chair Sutley (a former aide to Browner when she was EPA administrator).

Browner was a consummate Washington insider. Having gotten her start in Washington as a young staffer to Senator Al Gore, she then went back to her home state of Florida to become head of its Department of Environmental Protection. This put her in perfect position to be the Clinton-Gore administration's pick in 1992 to run the federal EPA, where she became the longest-serving administrator.

Like other high-level officials who served in the Clinton-Gore administration, Browner was again well-positioned in 2008 for a senior role in the new Obama administration. She became the first assistant to the president for energy and climate change. Known as the White House "energy czar," she surely had the president's ear.

One of Browner's primary roles in that capacity was to win congressional approval for the president's climate bill. To give up a key bargaining chip in those negotiations without getting something in return did not seem to fit with her reputation for being a tough negotiator. Perhaps the strategy was to make an overture to the Republicans with the expectation that it would be reciprocated later.

The administration took a beating for announcing expanded offshore drilling without having in place the capacity to respond to a major blowout, but Browner, who reportedly had downplayed the risk of problems,[4] steadfastly defended the decision. She claimed that the Obama administration was actually cutting back on offshore drilling that had been in the five-year plan it had "inherited" from the Bush administration.[5]

THE OVAL OFFICE ADDRESS

The government's spin efforts steadily increased over time as the oil flowing in the Gulf remained unstoppable. On June 7, on the heels of his third trip to the region, the president met with his cabinet and then the press to address the disaster, promising that he would not let BP "nickel and dime" the people of the Gulf, and that they would get the help they needed.[6]

Prodded in a television interview aired that evening to respond to criticism that he was spending too much time talking to experts and not enough time in the Gulf, the president said he was talking to the experts so "I know whose ass to kick."[7]

On the evening of June 15, following a fourth trip to the Gulf, President Obama elevated his public relations campaign to new heights by using his first-ever Oval Office address to speak about the disaster. In terms of staging, the venue alone served to convey the gravity of the situation (Obama would use his second Oval Office address in late 2010 to announce the official termination of U.S. military action in Iraq, and his third in late 2015 to announce plans to combat domestic terrorism in the wake of the San Bernardino, California, shooting rampage).[8]

Sitting at the same ceremonial desk in the Oval Office that was used by his predecessors to speak of warfare, President Obama framed the oil disaster in the Gulf in similar terms. After making a reference in the opening paragraph of his speech to the fight against al Qaeda, Obama referred to "the battle we're waging against an oil spill that is assaulting our shores and our citizens." Later in the speech, he pledged to "fight this spill with everything we've got for as long it takes" and laid out the "battle plan."[9]

Obama told the American people that he had activated the National Guard. Some 17,000 members of the Louisiana National Guard were mobilized at the end of April. Their role was to help with a variety of tasks, including keeping the oil from coming ashore. A small Ohio Air Force Reserve squadron specializing in aerial spraying of insecticide for pest control was activated to spray dispersant on oil.

Early in the president's speech, there was another reference with military ties, if not obvious on the surface. Obama reported that he had "assembled a team of our nation's best scientists and engineers to tackle this challenge."[10] Leading the team would be Secretary of Energy Chu. The team included some of the "Jasons," a group of the nation's best and brightest scientists that was formed in the early years of the cold war to advise the government, primarily on military issues.[11]

One of the Jasons in the Gulf was Richard Garwin, who helped invent the hydrogen bomb and cap the burning oil wells after the second Gulf War.[12] Although Chu early on had ruled out any form of explosives, there were suggestions from such notable figures as former president Bill Clinton to literally blow up the well.[13]

In his speech, Obama also said that he had called on the secretary of the Navy, Ray Mabus, to develop a long-term restoration plan for the Gulf. The appointment of the secretary of the Navy to this task seemed unrelated to any military objective. Rather, as Obama put it, Mabus, a former governor of Mississippi, was qualified because he was a "son of the Gulf."

The son of the Gulf turned to a daughter of California, Kate Brandt, to make the plan happen. A 25-year-old political appointee at the time, Brandt had all of nine months' experience under her belt working in Carol Browner's office at the White House before she went to work at the Pentagon as an aide to Mabus. Brandt assembled and shepherded an ad hoc task force of representatives from an array of federal agencies. After consultation with various Gulf stakeholders, the task force produced an initial plan in under four months, an impressive accomplishment. The Gulf Coast Ecosystem Restoration Council subsequently took more than four years to come up with a full restoration plan.[14]

The warfare metaphor in the Oval Office address was a poor choice for speechwriters from a diplomatic perspective. The blowout, devastating as it was, had been an accident, not an intentional armed attack. Although most Americans might not have been aware that BP was a British company, most Britons were. Some began to wonder whether BP was being slapped around because of its heritage—notwithstanding the absence of any reference in the president's speech to the War of 1812 (when the British burned down the White House).

From one perspective, however, the warfare metaphor was apt. In addition to loss of life, warfare typically involves large-scale unintended and long-lasting devastating consequences for which nations are often grossly unprepared. The BP disaster would prove to be analogous to war, if only in that sense.

Using another metaphor, Obama made promises about curing the oil "epidemic." He predicted that "in the coming days and weeks . . . up to 90 percent of the oil that is leaking out of the well" would be captured.[15] In an editorial the next day, the *New York Times* seemed skeptical of the prediction and accused the president of being "less than frank about his administration's faltering efforts to manage this vast environmental and human disaster."[16]

Finally, Obama vowed in his speech to "make BP pay for the damage their company has caused."[17] This part of the battle cry was likely in anticipation of the White House meeting that had been scheduled the very next day with BP top brass to announce the establishment of its $20 billion trust fund to pay disaster expenses.

THE VICTORY CELEBRATION

The well was finally capped on July 15, 2010. The administration estimated that by then approximately 5 million barrels had been discharged from the well, at an average rate of 57,000 barrels a day (including oil that was captured at the wellhead). The administration waited a few more weeks to be sure the cap would hold. Then, it capped its spin strategy by exaggerating the amount of oil that had been captured.

On August 4 Carol Browner went on the morning television networks (six in all) to announce a "turning point" in the containment of the oil. She claimed that 75 percent of the oil had been eliminated thanks to the response efforts. "The containment, burning, skimming, it worked," she said, "and Mother Nature will continue to break it down."[18]

Browner's 75 percent figure was inaccurate. After hearing it that morning, at 8:45 A.M., NOAA Administrator Lubchenco fired off an email, protesting to high-level aides at the White House and Homeland Security:

I'm concerned to hear that the oil budget report is being por-trayed as saying that 75% of the oil is gone and that this is a NOAA report. Please help make sure that both errors are corrected: It's not accurate to say that 75% of the oil is gone. 50% of it is gone—either evaporated or burned, skimmed or recovered from the wellhead.[19]

Having made her objection, Lubchenco dutifully proceeded that afternoon to the White House to participate in the next piece of victory staging—a press room news conference. But duty only went so far for the renowned marine ecologist who was recruited by the Obama administration to be the first female administrator of NOAA. As Lubchenco waited offstage with White House press secretary Robert Gibbs, Browner, and Coast Guard admiral Thad Allen, Gibbs pushed her to be consistent with Browner's morning television appearances. In Lubchenco's words, he "nearly had a cow" when she told him that she wasn't going to back Browner's 75 percent figure.[20]

Although Lubchenco held firm on that score, she otherwise played the good soldier, presenting a pie chart depicting the dispo-sition of the nearly 5 million barrels of oil then estimated to have come out of BP's well. Gibbs tried to lighten things up by jokingly referring to himself as Vanna White from the television game show "Wheel of Fortune," as he served as Lubchenco's human pointer for the pie chart.[21]

After Lubchenco claimed that the calculations used for the pie chart data had been peer-reviewed, Browner seized on that com-ment to repeat "peer review, peer review, and peer review."[22] "Peer review" was a particularly loaded term for Browner to use consid-ering it was the very same one that earlier proved troublesome for her office when it was working on the moratorium on oil drilling in the Gulf. Although there is no evidence that either Lubchenco or Browner knew it at the time, the calculations had not been peer reviewed.

The press and even congressional Democrats jumped all over the press room act. Once again, the administration had to backtrack, and it wasn't long before the pie chart disappeared from the White House website.

Lubchenco later conceded that some of the figures in the pie chart were inaccurate and that it had not been peer reviewed.[23] She also did not dispute that one of NOAA's scientists had gotten the initial flow rate very wrong. His estimate was 5,000 barrels/day, when it was closer to 50,000.[24]

The Presidential Commission staffer (one of the panel's former Supreme Court law clerks) who investigated the various errors poignantly concluded: "By initially underestimating the amount of oil flow and then, at the end of the summer, appearing to underestimate the amount of oil remaining in the Gulf, the federal government created the impression that it was either not fully competent to handle the spill or not fully candid with the American people about the scope of the problem."[25]

Asked in retrospect about the government's public relations effort, Lubchenco said:

Everybody involved was intensely frustrated the media coverage was so critical of the federal government when we knew that everything was being done that could be done to fix the problem. Public perception was that the federal government didn't know what it was doing, and was blowing it. We saw a very different side of it. Everything wasn't perfect, but wasn't anything near as inept as the public was being led to believe. . . . No doubt the White House made mistakes. I made mistakes.[26]

It must have been a long summer for the administration's top officials involved in managing the disaster. This was likely particularly true for Browner (who declined to be interviewed for this book), who at the same time was managing the climate bill battle on Capitol Hill. By the time of the August 4 press conference, the climate bill had gone down in flames.

OBAMA'S KATRINA?

Midway into the BP disaster, an ABC News/*Washington Post* poll put public disapproval of the federal government's response at 69 percent and made comparisons to poor public perception of its response to Katrina.[27] Was the Obama administration's spin designed at least partly to keep the BP disaster from becoming perceived as "Obama's Katrina," as some critics (including Bush loyalist Karl Rove) were calling it at the time?[28] Hurricane Katrina struck the New Orleans area in late August 2005, killing more than 1,800 people. The storm highlighted not only the Bush administration's ill-preparedness for the event, but also its poor media skills.

There are probably many more differences than comparisons to be made between the two calamities, especially in terms of loss of human life. But one legitimate common theme is the federal government's flailing image control.

President Bush was at the end of a five-week "working vacation" at his Texas ranch when the storm hit. On August 31 he took a slight detour from his normal flight path from Waco, Texas, to Andrews Air Force Base to fly over the storm-struck region. The White House released a photo of the concerned president looking out of an Air Force One window.

On his next flight to the region on September 2, 2005, Bush's handlers scripted an Air Force One landing at a regional airport in Mobile, Alabama, more than 130 miles from New Orleans. There, Bush held a photo opportunity in an aircraft hangar with high-level federal and state officials, including Federal Emergency Management Agency Director Mike Brown, upon whom he heaped the now infamous praise: "Brownie, you're doing a heck of a job."[29]

While Bush was congratulating Brown, dead bodies were literally still floating around New Orleans, and tens of thousands of suddenly homeless people were still holed up in the New Orleans Convention Center. Ten days later, Bush replaced Brown in the relief effort with Coast Guard admiral Thad Allen, whose name would again become a household word after he was called in to lead the federal response effort during the BP disaster.

The Bush White House communications team was operating at a deficit at the time, because its 33-year-old director, Nicolle Devenish, was on Mykonos, a popular Greek island. She was there to marry a fellow Bush politico, Mark Wallace. The power couple's wedding date was September 3, 2005, five days after Katrina had struck. Other Bush heavyweights were also on hand in Mykonos for the wedding, including the president's 2004 reelection campaign chair Ken Mehlman, who by then was chair of the Republican Party. Interviewed later, Mehlman was quoted as saying about the storm: "I certainly recognized watching CNN that it was not good. . . . And it required a large infusion of ouzo [a Greek liquor]. Which I had every night. My answer to Hurricane Katrina was Hurricane Ouzo."[30]

If the Obama administration's spin response to the BP disaster floundered at times, it never reached the level of the Katrina debacle.

In early December 2010 the Obama administration reversed its March 31 announcement of expanded offshore oil drilling, pending further study. The next month, Carol Browner announced she was leaving the administration. By January 2013 she was forcefully opposing drilling in the Arctic: "Developers and President Barack Obama's administration assured us these operations would be safe, thanks to strict oversight and new technology. Now it seems this optimism was misplaced."[31]

In January 2015 the Obama administration announced once again that it would allow offshore drilling off the Atlantic coast, while imposing some limits in the Arctic. In March 2016, however, it reversed itself again, and abandoned offshore drilling in the Atlantic.

BP'S SPIN GAME

BP launched a major crisis management effort immediately following the accident. Like the government's, it backfired at times.

From the start and at his own initiative, CEO Tony Hayward became the face of the disaster for BP. He went onto the Louisiana beaches to show his concern. Television commercials depicted him telling the audience with his British accent that "BP will make things right." His efforts didn't play well to much of the American public, or, for that matter, to some members of the BP public relations team.[32]

But Hayward was the CEO, and who was going to tell him to stop? Hayward's comments went from bad to worse when, at the end of May, he lamented, "I'd like my life back," prompting mockery in the British press.[33] When he was spotted sailing on his yacht in a regatta off the coast of England in late June, the press vilified him further.

Not long after, he succeeded in getting his life back, at least in one sense. In late July he resigned under pressure. Despite his own best-intentioned efforts, the BP disaster had become Hayward's Texas City.

The spin that came back to haunt BP most involved its lowball flow rates. As discussed in chapter 4, through a web link from Capitol Hill, media outlets began providing their worldwide audiences with a live feed of the oil gushing into the Gulf. Once Spillcam's footage became public, experts began using it to try to calculate the flow rate. Their numbers did not match up with BP's. Federal investigators became convinced that BP was misrepresenting how much oil was flowing from the well in an effort to underplay the seriousness of the accident.

According to the government, BP underestimated the flow rate at 5,000 barrels a day even after its own engineers had reason to know that it was far higher.[34] BP, the government alleged, received estimates ranging anywhere from 14,000 barrels/day to 146,000 barrels/day.[35] The government claimed that in an e-mail to BP executives, one engineer had specifically cautioned against using the 5,000 barrels a day figure, saying the number could even be as high as 100,000 barrels a day (a figure that was within the 64,000–110,000 range initially estimated by the Coast Guard in its log on April 23, 2010).[36]

Despite these other estimates, BP stuck by the 5,000 barrels a day figure, including in written representations to a congressional subcommittee investigating the disaster. As a result, the company wound up being charged with obstructing Congress, a felony to which it pled guilty (as discussed in chapter 6).[37]

BP's television commercials continued once the well had been plugged that summer. But no more British accents. Instead, a BP employee with a Cajun name and accent held up a big fish and told the American public it was time to come back to the Gulf. In a later commercial, a high-level BP official in Alaska described BP's investment in America and American jobs—with no discernible accent.

Beginning in 2013, BP launched a major public relations campaign designed to portray itself as the victim, not villain, in the claims process (as discussed in chapter 5). This spin effort made use of major media advertising that pointed both to the asserted unfairness of the overall claims process as well as to specific instances in which awards had been made to unworthy claimants.

In the spring of 2015, timed around the fifth anniversary of the disaster, BP unveiled yet another television commercial that was heavily aired. Like its other advertising, this one featured an actual BP employee. "What happened here five years ago changed us," he proclaimed after speaking of toughened safety standards, increased training and monitoring, and greater authority for employees to stop jobs when they encounter unsafe practices. "Committed to America and a safer BP," read the message at the end.[38]

Perhaps so. Only time will tell.

10

MYTH V. REALITY

By the time BP appeared in U.S. District Court in New Orleans in January 2013 to plead guilty to criminal charges associated with the disaster in the Gulf, it was already a recidivist (a repeat offender). Its "rap sheet" listed numerous environmental crimes, including the deaths of workers killed in an explosion and fire at its Texas City refinery in 2005.

Chief Judge Sarah Vance took particular note of BP's record when she accepted its guily plea: "The BP family of companies has a history of deficient safety management, " she wrote. "If past is prologue, only a sentence several orders of magnitude more severe than any previously imposed on any BP company will be sufficient to achieve adequate deterrence."[1]

Despite this poor record, BP had managed to paint itself with a green brush, largely as a result of the efforts of its former CEO, Lord John Browne. When that image is compared to reality, it becomes apparent that BP's paint job was "greenwashing." After Tony Hayward took the helm at BP in 2007, the company made less effort to portray itself as envrionmentally conscious. Shortly after the disaster, BP began selling off its alternative energy assets to pay for disaster-related costs. Perhaps by then it had given up on ever being called a green leader again.

REALITY: BP'S RECORD

As noted earlier, the federal government has discretion to handle environmental violations as civil matters, criminal matters, or both. BP's oil operations in the United States—whether drilling, transporting, or refining—have frequently resulted in both civil and criminal enforcement actions. The BP family history of civil violations dates to 1990, when one of its tankers ran aground off the coast of Southern California. The accident resulted in the discharge of 400,000 gallons of oil over 60 square miles of ocean. The case was settled for $3.9 million, a substantial penalty at the time.[2]

In 1999 BP pled guilty to the unlawful disposal of hazardous substances on the North Slope of Alaska. BP contractors had illegally injected paint thinner and toxic solvents down the outer rim of oil wells over a period of three years. The company paid $500,000 in criminal penalties, and an additional $6.5 million in civil penalties. Under the terms of the criminal plea agreement, the company also was required to establish a $15 million environmental management system at all of BP's facilities in the United States and Gulf of Mexico that were engaged in the exploration, drilling, or production of oil.[3] In addition, BP pledged "to use best environmental practices to protect workers, the public, and the environment."[4]

In 2001, BP paid multimillion civil fines for Clean Air Act violations. It was forced to reduce air emissions from refineries in eight states.[5]

In March 2005 the fatal explosion took place at the BP refinery in Texas City, Texas. BP pled guilty to a felony violation of the Clean Air Act and paid criminal and civil penalties. A report by an independent panel chaired by former secretary of state and

Treasury secretary James Baker roundly criticized BP's lack of safety process.[6] Additional violations at the same refinery in 2009 led to further civil penalties.[7]

In March 2006 a BP pipeline on the North Slope of Alaska leaked nearly 5,000 barrels of crude oil, the worst discharge in three decades there. BP paid both criminal and civil penalties under the Clean Water Act.[8] As one observer noted, "All too often, BP's management culture appears to place undue emphasis on cost-cutting, while favoring rhetoric over reality."[9]

Four more minor pipeline discharges on the North Slope occurred in 2007–11. BP paid civil penalties for them in 2014, wiping the slate clean before the penalty phase of the trial for its violations in the Gulf.[10]

MYTH: THE BROWNE ERA

BP's reputation for being a "green" company is often attributed to Lord John Browne, Tony Hayward's predecessor as CEO. In 1997 at Stanford University, Lord Browne made a landmark announcement for the head of an oil company: it was time for the world to pay attention to climate change.

Lord Browne deserves credit for leading the industry in publicly acknowledging that climate change is a problem, especially with international negotiations on the Kyoto Protocol climate treaty then approaching. Nonetheless, his Stanford speech was at least mildly distorting on the science, long on generalities, and short on specific commitments. For example, in downplaying the role BP was playing in climate change, Browne claimed that "only a fraction of the total emissions [20 percent] come from the transportation sector."[11] The 20 percent estimate, hardly negligible enough to be coined a fraction, was well below authoritative scientific estimates at the time.[12] Moreover, transporation sector emissions were known at the time to be steadily increasing in the developed world, which bore the brunt of the responsibility for climate change.

Lord Browne's Stanford speech also contained no quantifiable goal to reduce BP's carbon emissions level. Rather, he pledged only vaguely to "monitor and control our own carbon emissions."[13] It was not until more than a year later, in 1998, that Browne actually set a reduction target. In a speech at Yale's Business School, he pledged to reduce BP's 1990 carbon emission levels by 10 percent by 2010.

Stanford researchers David Victor and Joshua House have closely and credibly scrutinized the plan Browne announced at Yale.[14] They found that BP set the 10 percent goal after undertaking an analysis of what reduction could be achieved by its business units at virtually no cost. To help it bring down carbon emissions, BP also utilized a novel intra-corporation emissions trading system under which no money changed hands (a tax advantage).

Victor and House report that BP experimented with a pilot of the system in the third quarter of 1998 using 12 business units and launched it full scale in 2000.[15] Since the 10 percent reduction goal was scheduled to be reached by 2010, attaining it over ten years represented an average reduction of only 1 percent a year.

This unambitious goal turned out to be even more easily reached than expected. The 10 percent reduction was largely achieved simply by increasing energy efficiency and cutting back on methane released from inefficient flaring and venting at oil wells. In fact, by the end of 2001, the entire 10 percent reduction had already been achieved, and BP suspended its emissions trading system.[16]

In his 1998 Yale speech, Lord Browne emphasized that BP's 10 percent reduction figure exceeded the Kyoto Protocol's 5 percent target for industrialized nations. A better comparison, however, would have been to DuPont's pledge to reduce its 1990 carbon emissions levels by 65 percent by 2010 (it succeeded in cutting them by 45 percent by 2000).

The modesty of BP's greenhouse gas reduction plan ended up being buried in a remarkably successful marketing campaign. In 2000, amid much fanfare, BP launched its new green and sunny yellow Helios logo and rebranded itself "Beyond Petroleum." Ogilvy, BP's public relations firm, received two separate awards (and one honorable mention) for the effort from *PR Week* in 2001. BP's public relations campaign earned it a very different kind of award in 2002. BP received the "Best Greenwash Award" from Friends of the Earth (the left-leaning environmental public interest group) at the World Summit on Sustainable Development in Johannesburg, South Africa. Friends of the Earth, joined by other organizations, lambasted BP for spending more on its rebranding campaign than on its solar investments.[17]

Whether a visionary or not, Lord Browne was a savvy businessman who likely saw value in his company's being perceived as green. It is unclear whether BP's initiatives were motivated primarily by an interest in enhancing its reputation, making the planet a better place, or raising its financial bottom line. It is clear that BP's green initiatives were markedly small scale and by no means representative of the company-wide culture. The highly touted emissions reduction plan that concluded in 2001 not only was unambitious and short-lived but lacked follow-up. In the subsequent four-year period culminating in 2005, BP's greenhouse gas emissions remained largely unchanged.[18] If BP had grown greener, it was by a very light shade of green.

THE HAYWARD ERA

Once Tony Hayward took over as CEO in 2007, the green image dimmed. Myth or reality, BP's green rebranding was Lord Browne's brainchild. With his hasty exit due to a personal scandal came another rebranding, although BP officially downplayed the difference between the two:

The new brand value, "Responsible," encompasses BP's original aspirations towards the environment, in addition to other key areas such as safety and social welfare. Our aspirations remain absolutely unchanged: no accidents, no harm to people and no damage to the environment.[19]

The environmental activist community was just as skeptical of Hayward's initiative as it had been of Lord Browne's. In 2008 Greenpeace gave BP its Emerald Paintbrush Award for greenwashing:

You wouldn't know it from their adverts, but BP bosses are pumping billions into their oil and gas business and investing peanuts in renewables. They've won the 2008 Emerald Paintbrush award because their slogans suggest that they are serious about clean energy, while their actions show they're still hell-bent on oil extraction.[20]

In 2009 Tony Hayward gave a speech at Stanford Business School that appeared to confirm that he had abandoned any remnants of Browne's interest in being perceived as green:

We had too many people that were working to save the world. We'd sort of lost track of the fact that our primary purpose in life is to create value for our shareholders. And we failed to realize that we're an operating company.[21]

The notion that a corporation's primary purpose is to increase shareholder value ("shareholder value theory") has been the subject of debate among scholars for years. Cornell law professor Lynn Stout calls the concept a myth. In Stout's view:

As long as boards do not use their power to enrich themselves, the law gives them a wide range of discretion to run public corporations with other goals in mind, including growing the firm, creating quality products, protecting employees, and serving the public interest. Chasing shareholder value is a managerial choice, not a legal requirement.[22]

Stout's book is full of examples drawn from BP. She maintains that BP shareholders do not necessarily want to raise share value to the exclusion of any other interest. "Real human beings own BP's shares, either directly or indirectly through pension and mutual funds, and real human beings care about much more than just whether BP stock rises."[23]

A more enlightened current view of a corporation's purpose is known as the stakeholder theory. It teaches that a corporation owes a duty not just to its shareholders but to all of its stakeholders. These stakeholders include its business partners, customers, employees, and communities, among others. As discussed in earlier chapters, many of BP's stakeholders were adversely affected by the blowout. They included BP's shareholders, whose stock plummeted. Tony Hayward's focus on being an "operating company" backfired from any perspective.

THE RANKINGS

The lack of substance behind BP's reputation as a green company largely escaped close scrutiny, perhaps thanks to the the efforts of John Browne and the effectiveness of BP's public relations campaign. As Frances Beinecke, former president of the Natural Resources Defense Council, put it, "We were all hoodwinked by John Browne."[24]

A telling barometer of the campaign's success was BP's 2006 ranking as the number one international environmental leader ("WaveRider") in *Green to Gold*, a book by environmental experts Dan Esty and Andrew Winston that is commonly assigned in university environmental courses.[25] "If BP is a waverider, the wave is one of toxic sludge," wrote one skeptical reader in in 2008.[26] He was apparently referring to BP's reliance on tar sands oil (known to produce higher greenhouse gas emissions than other oil sources) at its refinery in Alberta, Canada.

When the revised paperback edition of *Green to Gold* was published in 2009, the authors readily acknowledged BP's fall from grace.[27] "In the past few years, the company has had some serious breaches in safety and operations that have called into question its environmental and social leadership," they wrote.[28] How did BP manage to maintain its green image in environmental circles in the face of its appalling compliance record? Esty and Winston quoted "one knowledgeable observer" as attributing that success to the effort that BP made in building its reputation: "It was fascinating how much slack the environmental community cut BP. Their investment in being seen as good guys paid off handsomely. If ExxonMobil had done the same thing, there would have been hell to pay."[29]

What was the competition like for the number one spot? Patagonia is "arguably the most environmentally focused company in the world," Esty and Winston wrote.[30] Patagonia would have been an easy pick for the top of the list. It is truly a leader, and its reputation is well-deserved—it practices what it preaches. For example, around Black Friday (the big post-Thanksgiving sales day) in 2011, the company ran a well-publicized "Don't Buy this Jacket" campaign. In an effort to reduce the effects of conspicuous consumption on the environment, Patagonia was encouraging its customers to think about whether they really needed another jacket. Imagine BP running a "Don't Buy this Gas" campaign to encourage drivers to cut back on non-essential fuel usage.

BP's artificially inflated ranking raises questions about the utility of rankings in general. There are now more than 100 different types of "sustainability" rankings (often replacing more narrow environmental rankings), some of them novel. In 2013 Green-Biz launched a "Twitter Index" that it describes as "our exclusive ranking of which corporate sustainability execs have a strong voice on Twitter."[31]

Given reasonable differences over how to define sustainability itself, there certainly can be differences about what metrics to use in measuring it. This point has not been lost on the sustainability community: one respected group, SustainAbility, is ranking the rankings. With improved metrics in place, perhaps companies with records like BP will not be at the top of a list of environmental high performers. Going forward, improved rankings systems also could provide incentives for companies to put greater resources into genuine sustainability programs instead of marketing.

11

THE MOTHER OF ALL RESTORATIONS

"Restoration" is a term of art used broadly to describe the process of trying to repair environmental damage caused by pollution. The word is a bit of a misnomer in the sense that often it is not truly possible to put things back the way they were. Like the disaster itself, the restoration required by the BP blowout is unprecedented in size. Indeed, it has been called the "Mother of all Restorations."

Given the complexity of the natural resource damage assessment as well as preexisting, long-term environmental damage in the Gulf of Mexico, it would have been very challenging to litigate how much of the injury was due to BP's oil. As is commonly the case, BP's responsibility for restoration never did go to court but was hashed out for years at the bargaining table. In the end, more than $15 billion in funding from various settlements (primarily with BP) will go toward restoration in the Gulf.

"WE GOT LUCKY"

"We got lucky," said National Oceanic and Atmospheric Administration Administrator Jane Lubchenco.[1] As bad as it was, the damage could have been worse, and the need for restoration even greater, were it not for several factors. First, because the disaster occurred in late spring and ran into summer, the Gulf waters were relatively warm. The temperature and natural flow of the waters facilitated biodegradation of the oil. Second, loop currents that might have taken the oil around the southern coast of Florida and up the east coast broke off. Third, the BP oil was light crude oil, a type that dissolves and degrades more easily than the heavy crude oil that flowed from the *Exxon Valdez* after it hit a reef off Prince William Sound in Alaska in 1989.[2]

The damage also could have been worse had it not been for the efforts of various government agencies, public interest groups, and the private sector. For example, the Florida state government collaborated with the National Fish and Wildlife Foundation and FedEx to relocate turtle eggs from a beach on the west coast of the Florida Panhandle to the east coast of the state. The scientists collected the eggs and FedEx transported them across the state in a specially designed and cushioned vehicle. The foundation reported that its work with FedEx resulted in the transfer of 25,000 endangered sea turtle eggs.[3]

EARLY STUDY AND RESTORATION

BP made laudable early commitments to studying the damage to the Gulf and jumpstarting the restoration process. In May 2010 the company established the Gulf of Mexico Research Initiative, through which it pledged $500 million to fund independent scientific research over a ten-year period. As BP emphasized at the time, this contribution was entirely voluntary. BP would not receive credit for those funds toward its legal liability for the disaster.

In April 2011, on the first anniversary of the blowout, BP made a $1 billion deposit of "early restoration" funds to be used to address natural resource injury caused by the disaster. Unlike the Gulf research institute, this funding came about as a result of a formal Early Restoration Framework Agreement with the federal and Gulf state governments under which BP ultimately did receive credit toward its full natural resource damage liability.

Projects using the early restoration funding proceeded in phases and were subject to the approval of the federal and state government agencies serving as the natural resource "trustees." The first two phases announced in 2012 comprised 10 projects estimated to cost a total of $71 million. The third phase, proposed in 2013 and finalized in 2014, had 44 additional projects at a total estimated cost of $627 million. Thus, the cost of the average early restoration project was nearly $13 million.

One of the third-phase projects was both expensive and controversial. The trustees approved the expenditure of $58.5 million to partially fund construction of a lodge and conference center at Alabama's Gulf State Park. A nonprofit group that preferred to see the funds spent on projects more directly related to the restoration of natural resources sued to stop the funding of the center. In February 2016 a federal judge in Alabama agreed with them and halted the use of early restoration funding for the project.

THE LONG-TERM RESTORATION PROCESS

In April 2010 Roger Helm was working for the U.S. Fish and Wildlife Service (a sister agency of the Interior Department's Minerals Management Service) at its headquarters offices in suburban Washington, D.C. Helm, a biological ecologist, was very experienced in looking at the effects of major oil discharges and their aftermaths. After the *Exxon Valdez* accident in 1989, Helm had worked for an Exxon subcontractor monitoring the effects of the oil on the nearshore plants and animals of Kodiak Island and the Alaska Peninsula.

Helm was hired in 1991 by the Fish and Wildlife Service to lead a natural resource damage assessment (NRDA) involving the Montrose Chemical Corporation, the manufacturer of DDT. (The pesticide, widely used in the United States until it was banned in

1972, was considered responsible for birds such as the bald eagle becoming endangered species.) A NRDA is conducted to identify the injury from major pollution and develop a restoration plan. It is a long and laborious process often involving many different federal and state government agencies and offices. They inventory the damage and debate solutions both internally and with the responsible parties. Helm worked on some 40–50 such assessments, including most of the largest hazardous waste and oil cases. The BP case was unlike any other.

When BP's well blew out in the Gulf on April 20, 2010, Helm was attending a conference in Portland, Oregon. He got a call from a colleague at NOAA telling him about the blowout. After that, he said, he did not go to a single conference session. Instead, he just stayed in his hotel room to be on the phone with Interior Department staff in its headquarters, regional, and field offices. He and his colleagues began to put together a contingency plan to cope with the oil they feared would be discharged. When the *Deepwater Horizon* rig sank on April 22, he said he knew "things were going to get a lot worse."[4]

Work on the BP disaster NRDA began almost immediately, and Helm was part of it. The Department of Commerce, the parent of NOAA, was designated the lead federal government agency. The Environmental Protection Agency and the departments of Agriculture, Commerce, and Interior—along with the five Gulf state governments—were designated the NRDA trustees.

The BP disaster NRDA was on a scale never before undertaken, either in terms of the science or bureaucracy. A discharge of such a high volume of oil and at such a depth was unprecedented. The assessment of the oil's impact in the deep ocean was especially difficult to study, as was its effect on very complex and interrelated ecosystems of the Gulf. Oil that remained in the water column (between the ocean surface and floor) conceivably could be passed up the food chain from the smallest organisms to the fish that feed on them and each other. Just because the oil may impact one species, however, does not necessarily mean that it will impact others. The damage "may or may not percolate up the food chain," said Professor David Valentine of the University of California at Santa Barbara, who has extensively studied the effects of the disaster on the marine environment.[5]

The NRDA process was made even more complicated in the Gulf because it was no easy task to distinguish between the harm BP caused and preexisting injury. The injury caused by BP's oil was by no means the first the Gulf had suffered either from a man-made or natural disaster. Rather, the Gulf has been bearing the effects of both for decades. One of the issues in high-level settlement negotiations on restoration was whether it would go beyond harm specifically linked to BP's oil.

The Gulf, which sits directly in "Hurricane Alley," has been hit with more and more hurricanes of increasing power in recent years, likely as a result of global warming. In 2007, recognizing the heightened danger and following Hurricanes Ivan, Katrina, and Rita during the 2004–05 hurricane seasons, the Minerals Management Service issued new guidance on the design of oil platforms for the region.

Oil and gas drilling, along with other commercial activity such as shipping, also has impacted the Gulf severely over the years. Satellite imagery on any given day shows oil sheens on the water near wells all over the Gulf.

As a result of the damage from various sources, the Gulf has needed wholesale restoration for a very long time. For example, an area in the northern Gulf has oxygen levels so low that fish cannot survive. In 2015 NOAA measured this long-existing "dead zone" at nearly 6,500 square miles, an area larger than the state of Connecticut.

As Bob Bendick of the Nature Conservancy put it, "overall Gulf restoration is as ambitious as any on Earth."[6] In 2012, anticipating that it would be receiving restoration funding as a result of the BP disaster, Louisiana prepared an ambitious Comprehensive Master

Plan for restoring just its coastline. The state used a 50-year planning timeline and an upper-end budget of $50 billion. The projects it identified included structural protection, bank stabilization, oyster barrier reef, ridge restoration, shoreline protection, barrier island restoration, marsh creation, channel realignment, sediment diversion, and hydrologic restoration.

In February 2016, after years of study, the natural resource trustees for the BP disaster released their final, 1,000-page report.[7] In the portion containing the findings of the injury assessment, the panel summarized the wide-ranging ecological impacts of the disaster:

> The Deepwater Horizon incident resulted in injuries to: intertidal marsh habitats, including marsh plants and associated organisms; shoreline beaches and sediments and organisms that live on and in the sand and sediment; fish and shellfish and other invertebrates that live in the water; a wide range of bird species; floating Sargassum habitats offshore and submerged aquatic vegetation; deep-sea and nearshore ocean-bottom habitats, including rare, deep water corals; four species of threatened or endangered sea turtles that live in the Gulf of Mexico; and several species of dolphins and whales.[8]

The 2016 report proposed 13 types of restoration under four general categories: Restore and Conserve Habitat, Restore Water Quality, Replenish and Protect Living Coastal and Marine Resources, and Provide and Enhance Recreational Opportunities.[9] Some of the many planned projects already under way include barrier island restoration, oyster reef restoration, seagrass restoration, marsh creation, and sediment work.

THE RESTORATION FUNDING

As a result of the criminal and civil settlements with BP and Transocean (the owner of the *Deepwater Horizon*), a total of more than

$15 billion will become available over a period of 15 years for restoration (broadly defined) of the Gulf. The bulk of the funding is from three sizeable and separate pools of money: (1) $8 billion designated in the BP settlement as natural resource damage payments (including the initial voluntary $1 billion "down payment"); (2) $5.4 billion (of the $7.5 billion) in civil penalties paid by BP and Transocean; and (3) $2.54 billion in criminal penalties paid by BP. Each of the funds has different criteria for how it is to be used.

The first pool of restoration funds constitutes the largest share of the money BP is paying in its civil settlement with the federal government and the Gulf states ($8 billion of the $20 billion). The funds are specifically designated to restore Gulf natural resources harmed by the disaster. Most of these funds are largely earmarked by state, with Louisiana getting by far the biggest chunk at $5 billion.

The second and third funding sources—penalty payments—are highly unusual. Ordinarily, all criminal and civil penalties paid to the federal government in environmental cases go directly to the general U.S. Treasury. Under the landmark legislation known as the RESTORE Act passed by the U.S. Congress in 2012, 80 percent of the Clean Water Act civil penalties recovered in connection with the BP disaster instead will go to Gulf Coast recovery. Thus, $4.4 billion of the $5.5 billion designated as civil penalties in the BP settlement and $800 million of the $1 billion in civil penalties in the Transocean settlement will go into a Restoration Trust Fund. Separately, $2.54 billion will go to a designated Gulf Environmental Benefit Fund under the criminal plea agreement between the federal government and BP.

The $5.4 billion Restoration Trust Fund is to be used for environment and economic development (broadly defined) in the Gulf. Thus, monies from that fund can be used for environmental restoration, economic recovery projects, and tourism and seafood promotion in the five Gulf states. This means that the funds can be spent for purposes that do not necessarily have any direct relationship to the injury caused by BP's oil. For example, Mississippi announced in 2015 that it will use $15 million of its funding to upgrade broadband connectivity along its coast, in an effort to improve education and health care there.

The $2.54 billion Gulf Environmental Benefit Fund must be used to remedy harm caused by BP's oil. Fifty percent of the funds will go to Louisiana for barrier island restoration and river diversion projects to improve coastal habitat. The remainder of the funding is divided among the other four Gulf states, to be used for unspecified projects.

Another unusual feature of the criminal plea deal is that the Gulf Environmental Benefit Fund is controlled by an environmental public interest group, not a government entity. Whereas control of the restoration funding resulting from payments under the civil settlement is vested in various state or federal officials (as is common), control of the Gulf Environmental Benefit Fund is vested in the National Fish and Wildlife Foundation (NFWF). As Juliet Eilperin of the *Washington Post* wrote at the time, it was as if the relatively obscure D.C.-based organization had won the lottery.[10] In a single legal document, NFWF was chosen to administer an amount of funding that exceeded the total it had handled in its entire 28-year history.[11]

NFWF had the only ticket to the lottery. An unusual foundation in the sense that it was created by an act of Congress and its board of directors is nominated by the secretary of the Interior, NFWF enjoys a cozy relationship with its federal "partners" who provide it with grant money. It was in the right place at the right time, in terms of being both well-connected to and well-respected by the federal agencies involved in selecting it for this important mission. It also had formed a relationship with BP early in the disaster by administering funds that the company provided from its share of revenue from the sale of oil recovered from the well site.

NFWF appears to be an effective conduit for administering federal funds outside the normal constraints applicable to federal and state agencies. It has put in place a Gulf restoration grant-making apparatus and awarded well-vetted, high-value grants at light speed compared to the normal pace of government agencies. Through late 2015 NFWF had made awards of nearly $500 million for more than 70 restoration projects in the five Gulf states, with Louisiana projects designated for nearly $220 million of that amount.

The ecological damage from the 3.2 million barrels of BP oil that spewed uncaptured into the Gulf only added to longstanding degradation requiring repair. Some much-needed broad-scale restoration is finally taking place. As the healing of the known damage begins, the long-term effects of BP's oil remain unknown.

12

HAVE WE LEARNED
OR ONLY FAILED?

Despite the steep drop in oil prices in recent years, oil production in the Gulf of Mexico is expected to climb to a record of nearly 2 million barrels per day by the end of 2017. Is it only a matter of time before a disaster of BP magnitude strikes again?

The best answer is one a good lawyer often gives: "It depends." It depends partly on what is done to change the status quo that failed to prevent the BP disaster. The likelihood of another disaster is directly proportionate to what the oil industry, individual companies, and the federal government do to minimize the risk.

The BP disaster is a case study not only in corporate failure, but also in public policy failure. The disaster provides important lessons that can help shape public policy on offshore drilling and corporate behavior generally.

LESSONS FOR THE OIL INDUSTRY

The Presidential Commission that closely studied the disaster called for formal oil industry self-policing, with the creation of a private sector oversight body akin to what exists in the nuclear power industry.[1] The oil industry was not interested in following that model, but did create the Center for Offshore Safety in 2011 to serve as an informational clearinghouse of sorts. The center is a toothless far cry from an oversight organization.

The oil industry needs to do more on its own initiative to curb offshore oil accidents: the industry's fatality rate in U.S. waters historically has been four times higher per person-hours worked than the comparable rate in European waters.[2] Fatal explosions on oil rigs have continued to occur in the Gulf of Mexico on an average of every other year since 2010, fortunately with limited environmental impact so far. In 2012 an explosion and fire on an oil rig operated by Black Elk Energy off the Louisiana coast killed three workers and resulted in an oil discharge into the Gulf. In November 2015 the Justice Department brought manslaughter charges against Black Elk and one of its contractors.[3] In June 2016, shortly into trial, Black Elk settled a civil case against it brought by the families of the dead, injured workers, and contractors.

The industry has made important strides toward ensuring it will not get caught flatfooted if another major accident does occur. Oil companies have made a significant joint investment (reportedly $1 billion) to increase response capability. Ten major companies are members of the Marine Well Containment Company, which is said to have the equipment, manpower, and knowhow to address the next catastrophe in the Gulf. Although it was not one of the original four founders of the coalition, BP is now a member. The Marine Well Containment Company also assists oil companies in obtaining drilling permits. Sixteen lesser known oil companies (that claim to extract half of the oil and gas annually in the Gulf) are members of a similar consortium known as the Helix Well Containment Group.

PUBLIC POLICY LESSONS

The Minerals Management Service (MMS) was the federal agency responsible for regulating offshore drilling at the time of the blowout. It was part of the Department of the Interior, whose headquarters is in a massive Depression-era structure that sits on the edge of the National Mall in Washington, D.C. (The Mall is overseen by the Park Service, another Interior agency.)

At its core, the public policy failure that contributed to the disaster was inadequate oversight by MMS of offshore oil drilling. The oil industry all but controlled its regulator—a phenomenon known as capture. Given the laxity of government oversight, the BP disaster was an accident waiting to happen.

MMS, created in 1982, was the brainchild of James Watt, President Ronald Reagan's secretary of the Interior. Before being entrusted with safeguarding the nation's natural resources, Watt had been an industry lobbyist and president of the Mountain States Legal Foundation, a conservative free enterprise and individual property rights organization created by Joseph Coors (of Coors Brewery). Watt's primary vision for MMS was to promote—not regulate—mineral extraction.

Watt later became infamous, making *Time* magazine's list of Top 10 Worst Cabinet Members ever.[4] He wound up coming under considerable fire for selling out the nation's environmental interests to big energy, as well as for his personal insensitivity. He reportedly compared his critics to Nazis and communists and tried to ban the Beach Boys from playing in their traditional July 4 concert on the National Mall. Referring to members of a coal advisory commission, Watt reportedly said, "We have every kind of mixture you can have. . . . I have a black, I have a woman, two Jews and a cripple. And we have talent."[5]

After resigning under pressure in 1983, Watt resumed his lobbying career, using his influence in the Reagan administration in a way that earned him an indictment for perjury and obstruction of justice.[6] He was able to avoid prison in a misdemeanor plea deal.[7]

Although MMS had the responsibility both to regulate and to promote offshore drilling, those objectives represented an inherent conflict of interest.[8] Dating from the Watt days, that conflict was largely resolved in favor of facilitating, rather than regulating, offshore drilling. Whether in the Reagan administration or subsequent ones, MMS served as a light-handed regulator at best, and one with many challenges. In its final report on the BP disaster, the Presidential Commission concluded:

> Federal efforts to regulate the offshore oil and gas industry have suffered for years from cross-cutting purposes, pressure from political and industry interests, a deepening deficit of technical expertise, and severely inadequate resources available to the government agencies tasked with the leasing function and regulation.[9]

One reason for the regulatory failures, according to the commission, was what it described as the "dominant role" of the American Petroleum Institute (API), the oil industry's trade and lobby group, in industry standard-setting.[10] In other words, the industry had heavy influence in writing the rules governing it, and as a result the rules were too lenient. "The inadequacies of the resulting federal standards are evident in the decisions that led to the Macondo well blowout," the commission concluded.[11]

Even if adequate standards had been in place, MMS had profound budgetary and operational problems that precluded it from closely monitoring the industry. For example, the agency had only one inspector for every 54 rigs in the Gulf. As Presidential Commission member Frances Beinecke noted, for an inspector to even get out to a rig, he needed to ask the company helicopter for a lift.[12]

Some MMS employees overseeing Gulf rigs from its Lake Charles, Louisiana, office received favors from the industry that crossed ethical and legal boundaries. In an investigation concluded in March 2010, the Office of the Inspector General in the Interior Department found "a culture where the acceptance of gifts from oil and gas companies were widespread."[13] Another employee reportedly conducted four inspections of a company with which he was simultaneously negotiating for a job he later accepted.[14]

Without adequate regulatory standards in place, and without sufficient capacity for monitoring and enforcement, MMS conducted little close scrutiny of offshore drilling. Its oversight of BP's operation at the Macondo well was no exception. It took MMS just six weeks to approve BP's initial exploration plan for the well, and just four weeks for MMS to approve BP's permit application to drill. The relatively quick turnaround times suggest that a very complicated and potentially dangerous project received less than intensive review. Moreover, MMS waived a full environmental impact assessment of the effect of drilling, granting a "categorical exclusion" for the well. BP had the "green light."

Regulatory reform

The Presidential Commission called for sweeping reform of how the federal government regulates the oil industry. The commission proposed new laws and major bureaucratic changes to improve oversight of offshore drilling. The recommendations for the Department of the Interior included creating a new regulatory agency, an entirely new risk management–based approach to regulation, increased expertise, greater reliance on consultants, heightened focus on safety, better reporting, and improved standards. Recommendations for other agencies included upgrading the response capabilities of the Coast Guard and Environmental Protection Agency in the event of an accident.

Only some of the commission's recommendations were implemented. Moreover, change was slow and ineffective considering the magnitude of the disaster and the ongoing threat. The Presidential

Commission, which was disbanded after issuing its official report in January 2011, was so dissatisfied with the slow pace of reform that its seven original members reconstituted themselves on the second and third anniversaries of the disaster. In April 2012 and 2013, the former commissioners took it upon themselves to issue additional reports assessing the progress made since their January 2011 report.[15] In the 2012 report, they warned that

> the risks will only increase as drilling moves into deeper waters with harsher, less familiar environmental conditions. Delays in taking the necessary precautions threaten new disasters, and their occurrence could, in turn, seriously threaten the nation's energy security.[16]

In both reports, the former commissioners were most critical of Congress for failing to enact legislation. In 2012 they gave Congress a "D." In 2013 Congress got a "D+."

In October 2011 the Interior Department created two rebranded separate entities to perform the functions formerly assigned to MMS. The Bureau of Ocean Energy Management (BOEM) now issues drilling permits, and the Bureau of Safety and Environmental Enforcement (BSEE) is now charged with oversight.[17] Although BOEM has issued more permits than ever before, BSEE did not issue new well control regulations until April 2016, six years after the disaster. Those regulations will be phased in over several years. In July 2016 BOEM and BSEE finally announced more stringent safety and environmental regulations for Arctic Ocean drilling.[18]

The reorganization seems to be in part an exercise in putting old wine in new bottles. Notwithstanding its makeover, the Interior Department seemingly remains incapable of playing the role of effective regulator. A report by the Government Accountability Office, the investigatory wing of the U.S. Congress, which assessed reform in the Interior Department in 2012, found that

Interior continues to face challenges following its reorganization that may affect its ability to oversee oil and gas activities in the Gulf of Mexico. Specifically, Interior's capacity to identify and evaluate risk remains limited, raising questions about the effectiveness with which it allocates its oversight resources. . . . It also continues to face workforce planning challenges, including hiring, retaining, and training staff.[19]

When interviewed shortly after the fifth anniversary of the BP explosion in 2015, BSEE director Brian Salerno, a retired Navy admiral, said, "I'm not confident the problem is solved."[20] This comment is especially troubling given the 36 percent increase in deepwater drilling in the Gulf of Mexico during the preceding four-year period.[21]

In 2016 the Chemical Safety Board, a relatively obscure but respected independent government agency that investigated the disaster, also continued to find problems. The board warned: "A culture of minimal regulatory compliance continues to exist in the Gulf of Mexico and risk reduction continues to prove elusive."[22]

As these reviews suggest, effective government oversight has yet to materialize and Congress has not overcome the gridlock and mustered the political will necessary to tighten offshore drilling laws. There is little cause for optimism. Although various components of the Interior Department can be reorganized and rebranded to an extent, history has shown that the agency simply cannot be pitted against an industry as powerful as the oil industry.[23] As the Presidential Commission noted, it's not so much a question of lacking legal authority as adequate political backbone:

The root problem has instead been that political leaders within both the Executive Branch and Congress have failed to ensure that agency regulators have had the resources necessary to exercise that authority, including personnel and technical expertise, and, no less important, the political autonomy needed to overcome the powerful commercial interests that have opposed more stringent safety regulation.[24]

Interior versus Big Oil is a classic David and Goliath mismatch. The oil industry has vast resources compared with those allocated to the Interior Department. The industry can afford to hire the best lobbyists and engineers, and is not constrained by government wages and red tape. Interior is a technological backwater compared to the oil industry. Relying on the agency to oversee the oil industry is simply a leap of faith, absent far greater support from the president and Congress. Other options should be considered.

Imprisonment

One option sometimes proposed is greater use of incarceration to punish people responsible for accidents like the BP disaster. The thinking is that if oil company employees feared time behind bars, they would be more likely to act responsibly in preventing accidents.

Incarceration, however, is a quick-draw solution to punishing white-collar and common crime alike. The United States incarcerates nearly 1 percent of its population—a startlingly high figure for a Western democracy—and for lengthy periods of time. Criminal enforcement of insider trading laws has reached an unprecedented level in recent years, with record sentences being handed down to some Wall Street offenders. For decades, more than 50 percent of the inmates in federal prisons have been drug offenders. But, it is not at all clear that incarceration has made a dent in either type of trade.

There is even less reason to believe that a heightened threat of incarceration would serve as an effective deterrent (or the other classic purposes of punishment) in the context of offshore drilling. Prison is not the answer.

Reducing Reliance on Oil

Prominent environmentalists and journalists alike have linked the BP disaster to Americans' dependence on oil. Peter Lehner, the former executive director of the Natural Resources Defense Council, has referred to "our oil addiction."[25] Jeremy Warner, business columnist for *The Telegraph,* wrote that "the environmental ruin now being visited on the Gulf of Mexico is not primarily about safety failures at BP, still less is it about lax regulation. Rather it is to do with America's insatiable appetite for oil."[26]

Blaming the disaster on an oil addiction is fair—to a point. It's a bit like blaming the drug violence in Mexico on American addicts. Drug demand drives drug violence. Oil demand drives oil drilling. On that theory, Americans potentially could decrease the risk of future disasters by decreasing oil consumption.

There are many ways to encourage Americans to curtail their consumption of oil, some more politically practical than others. Increasing automobile fuel efficiency through more stringent corporate average fuel economy standards has been very effective. Increasing gasoline taxes has never been popular. Radical approaches such as limiting the number of cars that can be registered in high-pollution cities (as in Beijing) or limiting automobile usage to every other day on high-pollution days (as in Paris) are politically untenable.

Americans should use less oil for many reasons. But that should be a separate conversation from the one about how oil companies can meet the demand for oil in a sustainable manner.

Reducing Offshore Drilling

Rather than increasing offshore oil production in the Gulf, the United States should consider decreasing it incrementally over time, at least in deep water. The debate over offshore drilling has been going on for decades, sparked by the 1969 Santa Barbara blowout. That incident led to long-standing federal and state bans on new offshore drilling off the California coast. Only a limited number of preexisting well sites there remain operational. After much debate, similar bans remain in effect along the East Coast. The rules for drilling in the Arctic are now tighter than ever.

Popular support for offshore drilling has declined in recent years—from 63 percent of Americans in early 2010 to 52 percent in 2014—and could continue to fall with the dramatic increase in oil supply.[27] In 2015 the United States produced more oil than it had in decades, largely as a result of hydraulic fracturing ("fracking"). Like it or not, the technology has resulted in recent years in what some consider an energy "revolution." It has been a game-changer in energy self-sufficiency and security; the United States now competes with Saudi Arabia as the leading oil producer in the world.

The safety of offshore drilling is relative. Ultimately, society defines what level of risk it is willing to bear, weighing the costs versus the benefits. And, ultimately, society imposes the terms upon which the risk can be taken. Given the increasing availability of onshore oil, natural gas, and renewable energy, society might soon be at its tipping point in terms of favoring reducing offshore drilling and the risk it brings.

Assuming such a tilt, it may just be a matter of time before politicians follow the will of their constituents, notwithstanding the disproportionate power of the oil lobby. It is time for the United States to start thinking about reducing, not increasing, offshore drilling, especially in deep water.

Reducing Drilling Privileges

Another option is to reduce the amount of offshore drilling performed by oil companies with poor records. Just as people can lose privileges when they abuse them, so should companies.

Compare, for example, one of the most fundamental privileges that most Americans take for granted: a driver's license. When the privilege to drive is abused, it is terminated, either temporarily or permanently. Not just drug and alcohol impairment, but even relatively minor motor vehicle offenses such as the failure to pay a speeding ticket can result in the suspension of a license. Or, at a much more serious level, compare the "three-strikes" laws in effect in many states. Repeat criminal offenders can lose the privilege of living in society—that is, be imprisoned for life—after committing three serious felonies.

Why should the privilege of extracting minerals in the ocean be treated any differently? Why not suspend a company's permit to drill for oil when the privilege has been abused, and ultimately revoke it in the case of repeated violations? Permission to operate could be withheld or withdrawn for a single well, the leasehold, a particular region, deepwater drilling, or all of American waters. As Presidential Commission co-chair and former senator Bob Graham commented, "It doesn't make walking around sense" to grant a deepwater drilling permit in the first place to a company with a bad record.[28]

Permanent revocation could mirror the three-strikes laws that apply to serious recidivists. Just as when a common criminal pays the ultimate price (short of death) after three serious felonies, an oil company that has committed three serious environmental felonies could be permanently banned from operating, if not in all of the country's waters, at least in the region where the felonies have occurred.

Such treatment fulfills the three classic purposes of punishment taught in criminological theory: deterrence, incapacitation, and just desserts. To apply the three theories: (1) if a company knew that it could lose its permit to operate if its conduct was sufficiently bad, it would have a greater incentive to act responsibly; (2) suspending or revoking a permit is the functional equivalent of incapacitating the company—that is, it cannot do further damage if it is not operating a well; and (3) a company that has acted egregiously in the pursuit of money deserves to lose money.

BP seems to have perceived accidents as a *cost of doing business*. Given its massive profits, that strategy historically might have made financial sense in all but the worst case scenario, such as resulted from the blowout at Macondo. With the recent decline in oil prices, oil companies can less afford to pay the extraordinary costs of a large-scale accident. Thus, there is added incentive to prevent one. Just imagine the still greater incentive that would be provided if companies were to associate accidents with a *loss of business*. In 2014 BP reported extracting the equivalent of 252,000 barrels of oil daily in the deepwater Gulf.[29] With oil averaging $93 per barrel that year, a company producing at that level would risk losing roughly $23.5 million per day in gross revenue there. It's hard to think of a greater deterrent, even at lower oil prices.

Such an approach also tracks free market principles. If a transgressing company were to lose its privilege to operate in a given region and forfeit its leases, other companies could move in. This scenario serves to instill a competitive advantage to companies that operate responsibly.

The denial, suspension, or revocation of drilling permits does not need congressional legislation or approval (although it could well meet with congressional resistance). The president could use his executive authority to instruct the Interior Department to issue and enforce regulations governing such an enforcement program, with the assistance of the Justice Department. The success of such a program would depend largely on strong White House support.

LESSONS IN CORPORATE LEADERSHIP, RISK MANAGEMENT, AND SUSTAINABILITY

A formula (albeit not a magical one) for companies to avoid crises like the BP disaster combines principled leadership, state-of-the-art risk management, and progressive sustainability practices. Companies also need to honor their responsibilities to their stockholders and other stakeholders.

As used here, principled leadership refers to setting priorities that transcend the balance sheet. Leaders should not permit the profit motive to be viewed as a trump card, especially over considerations such as compliance, safety, and environmental protection. Rather, they should establish a values-based corporate culture and ensure that it takes root at the operational level.[30]

At its best, risk management entails proactively identifying potential vulnerabilities and putting in place programs to minimize problems. Taking financial risk is an inherent part of business—but taking unnecessary operational risk is not. Moreover, unnecessary operational risks can result in undesirable and expensive consequences. The implementation of best practices and rigorous safety protocols is a good starting point.

Adherence to sustainable business practices is also an inherent part of doing business these days, and that means following fundamental tenets of social, environmental, and economic sustainability.

Social sustainability involves taking account of human impacts in making business decisions. This includes following such practices as fair wages and safe workplaces for employees, honoring civil and human rights, charging and paying reasonable prices, being honest with customers, respecting local communities, and being mindful of people's health.

Environmental sustainability involves taking account of ecological impacts in making business decisions. This includes protecting the air, water, land, oceans, natural resources, plant life, and wildlife. Shared resources must not be harmed.

Economic sustainability refers to how a company's actions affect its financial bottom line. Being economically sustainable means ensuring the company's long-term viability in making business decisions. This includes taking account of a wide range of financial impacts of decisions, including worst case scenarios.

The BP disaster teaches us that companies should be sustainable not just because it's the right thing to do morally, but also because it's just good business. A corporation need not choose between being sustainable and being profitable. It can and should be both.

Companies that embrace sustainability are more likely to prosper than companies that forgo it. They do not need to undergo a conversion and adopt Patagonia's vision of saving the world. They simply need to make decisions that are good from both sustainability and business standpoints.

The time has passed to view sustainability as a short-term fad or an unnecessary expense. A new generation of potential business leaders has grown up with sustainability, just as it has grown up with technology. Like technology, sustainability is here to stay, and provides great opportunity.

Corporations should reject the antiquated shareholder value myth in favor of the more enlightened stakeholder theory. They should acknowledge their responsibilities not only to their shareholders, but also to other stakeholders, such as their employees, contractors, business partners, and communities, among others.

The nation's worst sustainability disaster provides important lessons for the oil industry, the federal government, and business in general. The oil industry should do more to prevent offshore accidents. The federal government should consider reducing offshore drilling and put more teeth into regulation and enforcement. Business should have principled leadership, state-of-the-art risk management, and sustainable business practices, as well as an enlightened view of its responsibility to society. The BP disaster serves as an enormous wake-up call for the private and public sectors alike.

NOTES

CHAPTER 1

1. *In re: Oil Spill by the Oil Rig "Deepwater Horizon" in the Gulf of Mexico, on April 20, 2010*, United States District Court, Eastern District of Louisiana, Transcript of Nonjury Trial Proceedings Heard before the Honorable Carl J. Barbier, United States District Judge, Fleytas Deposition Exhibit 4472, Introduced at Phase I Trial on April 4, 2013 (www.mdl2179trialdocs.com/releases/release201304041200022/Fleytas_Andrea-Depo_Bundle.zip).

2. Remarks by the President to the Nation on the BP Oil Spill, White House, June 15, 2010 (www.whitehouse.gov/the-press-office/remarks-president-nation-bp-oil-spill).

3. Personal interview with Bob Bea, June 17, 2014.

4. *In re: Oil Spill by the Oil Rig "Deepwater Horizon" in the Gulf of Mexico, on April 20, 2010*, Findings of Fact and Conclusions of Law, Phase One Trial, 21 F. Supp. 3d 657, 674 (E.D. La. 2014, September 9, 2014, pp. 114, 129 (hereafter, "Phase One Decision") (www.uscourts.gov/courts/laed/9092014RevisedFindingsofFactandConclusionsofLaw.pdf). The underlying facts described here are based primarily on the court's decision.

5. Summarized from "*Deepwater Horizon* Oil Spill Final Programmatic Damage Assessment and Restoration Plan and Final Programmatic Environmental Impact Statement," Section 1.5.2. "Key Findings of the Injury Assessment," National Oceanic and Atmospheric Administration and other agencies, February 2016 (www.gulfspillrestoration.noaa.gov/sites/default/files/wp-content/uploads/Front-Matter-and-Chapter-1_Introduction-and-Executive-Summary_508.pdf) (hereafter, "Trustees Report").

6. Ibid.

7. National Commission on the BP Deepwater Horizon Oil Spill and Offshore Drilling, "Decision-Making within the Unified Command," Staff Working Paper No. 2, p. 15 (http://cybercemetery.unt.edu/archive/oil-spill/20121211010432/http://www.oilspillcommission.gov/sites/default/files/documents/Updated%20Unified%20Command%20Working%20Paper.pdf) (hereafter, the commission will be referred to as "Presidential Commission").

8. See www.whitehouse.gov/the-press-office/press-briefing-bp-oil-spill-gulf-coast.

9. See David Biello, "How Science Stopped BP's Gulf of Mexico Oil Spill," *Scientific American,* April 19, 2011 (www.scientificamerican.com/article/how-science-stopped-bp-gulf-of-mexico-oil-spill/).

10. Heidi Avery, "The Ongoing Administration-Wide Response to the Deepwater BP Oil Spill," White House Briefing Room, May 5, 2010 (www.whitehouse.gov/blog/2010/05/05/ongoing-administration-wide-response-deepwater-bp-oil-spill).

11. Tom Bergin, *Spills and Spin: The Inside Story of BP* (New York: Random House, 2011), p. 159; and Ed Crooks, "BP: The Inside Story," *FT Magazine,* July 2, 2010 (www.ft.com/cms/s/2/4e228e56-84ae-11df-9cbb-00144feabdc0.html).

12. Brian Montopoli, "Obama: Malia Asked 'Did You Plug the Hole Yet, Daddy?,'" CBS News, May 28, 2010 (www.cbsnews.com/news/obama-malia-asked-did-you-plug-the-hole-yet-daddy).

13. Personal interview with Jane Lubchenco, January 3, 2015.

14. "BP Boss Admits Job on the Line over Gulf Oil Spill," *The Guardian,* May 13, 2010 (www.theguardian.com/business/2010/may/13/bp-boss-admits-mistakes-gulf-oil-spill).

15. "BP Spills Coffee: A PARODY by UCB Comedy," June 9, 2010 (www.youtube.com/watch?v=2AAa0gd7ClM).

16. Personal interview, confidential source, November 6, 2014.

17. Ibid.

18. Ibid.

19. Martin Vander Weyer, "BP's Been Punished Enough—but Not Because Americans Hate the Brits," *The Spectator,* September 13, 2014 (www.spectator.co.uk/2014/09/bps-been-punished-enough-but-not-because-americans-hate-the-brits/).

20. Another reason for Feinberg's lack of popularity was his statement that the $20 billion victims fund set up by BP in the aftermath of the disaster was "an aberration." See www.nola.com/business/index.ssf/2015/04/feinberg_bp_oil_spill_fund.html.

21. Press Release, U.S. Attorney's Office, Northern District of Florida. January 14, 2014 (www.justice.gov/usao-ndfl/pr/five-sentenced-filing-fraudulent-bp-oil-spill-claims).

22. Press Release, U.S. Attorney's Office, Northern District of Florida. July 23, 2014 (www.justice.gov/usao-ndfl/pr/nine-sentenced-filing-fraudulent-bp-oil-spill-claims).

23. *United States v. Montgomery*, (No. 13-2596) (6th Cir. 2014) (unpublished opinion).

24. BP separately settled with the families of the deceased rig workers for undisclosed amounts.

25. *United States v. BP Exploration and Production, Inc.*, Reasons for Accepting Plea Agreement, No. 12-292 (E.D. La.) (January 29, 2013). "BP" is used in this book as Judge Vance used it: to refer generally both to the BP parent company and its subsidiaries.

26. World Economic Forum, Strategic Partners (www.weforum.org/about/strategic-partners).

27. "Sky-High Davos Summit Fees Leave Multinationals Feeling Deflated," *Financial Times*, October 9, 2014.

28. Kamal Ahmed, "BP Boss Says Deepwater Liabilities Exceed Cost of Hurricane Katrina," BBC News, January 22, 2015 (www.bbc.com/news/business-30933698).

29. See www.ncdc.noaa.gov/billions/events for the costs of Katrina.

30. "BP Second Quarter 2016 Results," BP, July 26, 2016 (www.bp.com/en/global/corporate/press/press-relesaes/second-quarter-2016-results.html). See also Michael Amon and Tapan Panchal, "BP Puts Tab for Gulf Disaster at $62 Billion," *Wall Street Journal,* July 14, 2016.

31. Fukushima (Japan, 2011) and Katrina (U.S., 2005) were natural (that is, not manmade) disasters. Chernobyl (Ukraine, 1986) was a government (that is, not corporate) disaster. Bophal (India, 1984) was a manmade corporate disaster, for which Union Carbide paid $470 million in compensation.

CHAPTER 2

1. See Santa Barbara Wildlife Care Network, quoting Fred L. Hartley, president of Union Oil Co. "I don't like to call it a disaster," because there has been no loss of human life. "I am amazed at the publicity for the loss of a few birds" (www2.bren.ucsb.edu/~dhardy/1969_Santa_Barbara_Oil_Spill/About.html).

2. *In re: Oil Spill by the Oil Rig "Deepwater Horizon" in the Gulf of Mexico, on April 20, 2010*, Findings of Fact and Conclusions of Law, Phase One Trial, 21 F. Supp. 3d 657, 730 (E.D. La. 2014) (www.uscourts.gov/courts/laed/9092014 RevisedFindingsofFactandConclusionsofLaw.pdf) (hereafter, "Phase One Decision").

3. Phase One Decision, p. 674.

4. Ibid., pp. 674–75.

5. Ibid.

6. Alison Fitzgerald and Stanley Reed, *In Too Deep: BP and the Drilling Race That Brought It Down* (New York: Bloomberg Press, 2011).

7. *In re: Oil Spill by the Oil Rig Deepwater Horizon in the Gulf of Mexico on April 20, 2010*, United States District Court, Eastern District of Louisiana, Transcript of Nonjury Trial Proceedings Heard before the Honorable Carl J. Barbier, United States District Judge, February 25, 2012 (Day One, morning), p. 76 (www.mdl2179trialdocs.com/releases/release201302250700001/2013-02-25_BP_Trial_Day_01_AM-Final.pdf) (hereafter "Trial transcript, Phase One, Day One") and Trial Exhibit 1694.

8. Phase One Decision, p. 685.

9. Ibid., p. 705 (language was in bold in court's original opinion).

10. Ibid., p. 701.

11. Ibid.

12. Ibid., p. 699.

13. Ibid.

14. *On Scene Coordinator Report Deepwater Horizon Oil Spill*, submitted to the National Response Team, September 2011, p. 182 (www.uscg.mil/foia/docs/dwh/fosc_dwh_report.pdf) (hereafter "OSC Report, September 2011").

15. Trial testimony, Phase 3, Day 1, January 20, 2015, p. 87 (www.mdl2179trialdocs.com/releases/release201501201000001/2015-01-20_AM_Opening_and_Adm_Austin.pdf).

16. OSC Report, September 2011, p. xiv.

17. Presidential Commission Report, chap. 5.

18. Personal/confidential source, November 3, 2014.

19. Personal interview with Garret Graves, May 18, 2015.

20. "NOAA Researchers Release Study on Emissions from BP/Deepwater Horizon Controlled Burns" (www.noaanews.noaa.gov/stories2011/20110920_gulfplume.html).

21. See "BP Spills Coffee: A PARODY by UCB Comedy," June 9, 2010 (www.youtube.com/watch?v=2AAa0gd7ClM).

22. Kenneth M. Pollack, *The Persian Puzzle* (Random House, 2004), p. 54.

23. Ibid., p. 52.

24. Ibid.

25. Personal interview with Bob Bea, June 17, 2014.

26. Ibid.

27. Ibid.

28. Ibid.

29. Personal interview, confidential source, September 2, 2014.

30. Email from confidential source, August 11, 2016.

31. Presidential Commission Report, p. 223.

32. See "Markey Statement on BP Spill Commission Hearing," Select Committee on Energy Independence and Global Warming, November 8, 2010 (www.markey. senate.gov/GlobalWarming/mediacenter/pressreleases_2008_id=0334.html).

33. Presidential Commission Report, p. 89.

34. BP Exploration and Production, Initial Exploration Plan, Mississippi Canyon Block 252.

35. Presidential Commission Report, p. 84 (noting this type of cookie-cutter plan was not at all unique to BP).

36. Personal interview with William Reilly, November 20, 2014.

37. Ibid.

CHAPTER 3

1. Summarized from Section 1.5.2 of Trustees Report.

2. Ibid.

3. Personal interview with Jane Lubchenco, January 3, 2015.

4. Ibid.

5. See http://sero.nmfs.noaa.gov/deepwater_horizon/size_percent_closure/ index.html.

6. Presidential Commission Report, p. 187.

7. See www.fws.gov/home/dhoilspill/pdfs/ConsolidatedWildlifeTable 042011.pdf.

8. National Wildlife Federation, "Five Years and Counting: Gulf Wildlife in the Aftermath of the Deepwater Horizon Disaster," March 30, 2015 (www.nwf. org/~/media/PDFs/water/2015/Gulf-Wildlife-In-the-Aftermath-of-the-Deep water-Horizon-Disaster_Five-Years-and-Counting.pdf).

9. Ibid.

10. Ibid. Additional NWF findings included discoveries of oil and dispersant compounds in the eggs of white pelicans nesting in three states (Minnesota, Iowa, and Illinois); less frequent spawning of spotted sea trout in 2011 in two states (Louisiana and Mississippi); dramatically lower numbers of juvenile red snapper; abnormal development in many fish species; damage to coral colonies in five separate locations; and oil sediment in a 1,200-square-mile radius of the wellhead.

11. Sabrina Canfield, "Judge Enjoins BP's Unconscionable Contract with Fishermen-Volunteers," Courthouse News Service, May 4, 2010 (www.court housenews.com/2010/05/04/26953.htm).

12. Ibid.

13. Personal interview with Gina Solomon, June 17, 2014.

14. "Deadly Dispersants in the Gulf: Are Public Health and Environmental Tragedies the New Norm for Oil Spill Cleanups?," Executive Summary, Whistle blower.org, undated (www.whistleblower.org/sites/default/files/Executive_Summary_Corexit.pdf).

15. OSHA Fact Sheet, "Current Training Requirements for the Gulf Oil Spill," July 21, 2010 (www.osha.gov/oilspills/Basic_Training_Fact_07_02_10.pdf).

16. Press Release, U.S. Department of Justice, January 24, 2013 (www.justice.gov/opa/pr/individual-pleads-guilty-id-fraud-and-impersonating-osha-official-wake-gulf-oil-spill).

17. See Hari M. Osofsky, Kate Baxter-Kauf, Bradley Hammer, Ann Mailander, and Brett Mares, "Environmental Justice and the BP *Deepwater Horizon* Oil Spill," *NYU Environmental Law Journal* 20 (2011), pp. 99, 182.

18. Ibid., pp. 184–86.

19. See Ian Urbina, "Banned Trailers Return for Latest Gulf Disaster," *New York Times,* June 30, 2010.

20. Michael Robichaux, quoted in "Louisiana Doctor Describes Clusters of Ailments among Gulf Residents," CNN, April 21, 2001 (www.cnn.com/2011/HEALTH/04/20/gulf.oil.illnesses/).

21. Bernard Goldstein, Howard Osofsky, and Maureen Lichtyeld, "The Gulf Oil Spill," *New England Journal of Medicine* 364 (April 7, 2011), pp. 1334–48 (DOI: 10.1056/NEJMra1007197).

22. J. H. Diaz, "The Legacy of the Gulf Oil Spill: Analyzing Acute Public Health Effects and Predicting Chronic Ones in Louisiana," *American Journal of Disaster Medicine* 6, no. 1 (January-February 2011), pp. 5–22 (www.ncbi.nlm.nih.gov/pubmed/21466025).

23. Mark D'Andrea and G. Kesava Reddy, "Health Consequences among Subjects Involved in Gulf Oil Spill Clean-up Activities," *American Journal of Medicine* 26, no. 11 (November 2013), pp. 966–74 (DOI: http://dx.doi.org/10.1016/j.amjmed.2013.05.014).

24. L. M. Grattan, S. Roberts, W. T. Mahan, P. K. McLaughlin, W. S. Otwell, and J. G. Morris, "The Early Psychological Impacts of the Deepwater Horizon Oil Spill on Florida and Alabama Communities," *Environmental Health Perspectives* 119, no. 6 (2011), pp. 838–43.

25. David M. Abramson and others, "Children's Health after the Oil Spill: A Four-State Study Findings from the Gulf Coast Population Impact (GCPI) Project," National Center for Disaster Preparedness Briefing Report, Columbia University, 2013 (http://academiccommons.columbia.edu/catalog/ac%3A156715).

26. Tonya Cross Hansel, Howard J. Osofsky, Joy D. Osofsky, and Anthony Speier, "Longer-Term Mental and Behavioral Effects of the Deepwater Horizon Gulf Oil Spill," *Journal of Marine Science and Engineering* 3 (October 20, 2015), pp. 1260–71.

27. "Five Years Later—What Have We Accomplished?," GuLF Study, 2015 Newsletter (https://gulfstudy.nih.gov/en/GuLF_2015_Newsletter_Final_508.pdf).

28. "Fisherman Files Restraining Order against BP," CNN, May 31, 2010.

29. Richard J. L. Heron, "Response to 'The Unknown: Respiratory Effects of Cleaning up an Oil Spill,'" *The Lancet* (Respiratory Medicine) 1, no. 5 (July 2013) (DOI: http://dx.doi.org/10.1016/S2213-2600(13)70127-9).

30. Bob Weinhold, "Emergency Responder Health: What Have We Learned from Past Disasters?," *Environmental Health Perspectives* 118, no. 8 (August 2010), pp. A346–A350.

31. Ibid.

32. "Oil Spill Dispersant [COREXIT EC9500A and EC9527A) Information for Health Professionals," Centers for Disease Control and Prevention, Agency for Toxic Substances and Disease Registry, May 13, 2010.

33. Ibid.

34. "Statement by EPA Administrator Lisa P. Jackson from Press Conference on Dispersant Use in the Gulf of Mexico with US Coast Guard Rear Admiral Landry," Environmental Protection Agency, May 24, 2010 (https://archive.epa.gov/bpspill/web/pdf/statement-dispersant-use-may24.pdf).

35. See Hari Osofsky and others, "Environmental Justice and the BP *Deepwater Horizon* Oil Spill," p. 99 (and sources cited therein), upon which I have relied both for general themes and specific data.

36. Robert D. Bullard, *Dumping in Dixie* (Boulder, Colo.: Westview Press, 1990), ch. 2.

37. The most recent striking example of an environmental justice issue arose in Flint, Michigan, in January 2016, when it came to light that its local drinking water was contaminated with lead. Some 18 months earlier, Flint switched from the Detroit water supply to the Flint River as its source. Because the river water was more acidic, it had a corrosive effect on the old pipes carrying it, resulting in highly unsafe lead levels in the drinking water. Flint is the second poorest city of its size in the country, and the majority of its residents are black.

38. Environmental Justice, Environmental Protection Agency (www3.epa.gov/environmentaljustice). According to EPA, environmental justice is "the fair treatment and meaningful involvement of all people regardless of race, color, national origin, or income with respect to the development, implementation, and enforcement of environmental laws, regulations, and policies."

39. Hari Osofsky and others, p. 168. Also see Catherine Clifford, "BP Hires the Unemployed for Clean-Up," June 8, 2010, CNN.com (http://money.cnn.com/2010/06/08/smallbusiness/bp_hiring_unemployed/).

40. Abe Louise Young, "BP Hires Prison Labor to Clean Up Spill While Coastal Residents Struggle," *The Nation*, July 21, 2010.

41. Robert Bullard, quoted in Perry E. Wallace, "Environmental Justice and the BP Oil Spill: Does Anyone Care about the 'Small People' of Color?" (www.wcl.american.edu/environment/WallaceVolume6Issue2.pdf).

CHAPTER 4

1. Statement by the Press Secretary on the President's Oval Office Meeting to Discuss the Situation in the Gulf of Mexico, White House, April 22, 2010 (www.whitehouse.gov/the-press-office/statement-press-secretary-presidents-oval-office-meeting-discuss-situation-gulf-mex).

2. White House, *Management of Domestic Incidents*, Homeland Security Presidential Directive-5, Washington, D.C., February 28, 2003 (www.fas.org/irp/offdocs/nspd/hspd-5.html).

3. Heidi Avery, "The Ongoing Administration-Wide Response to the Deepwater BP Oil Spill," White House Briefing Room, May 5, 2010 (www.whitehouse.gov/blog/2010/05/05/ongoing-administration-wide-response-deepwater-bp-oil-spill).

4. Ibid.

5. Personal interview with Jane Lubchenco, January 3, 2015.

6. Personal interview with William Reilly, November 20, 2014.

7. Ibid.

8. Ibid.

9. Jane Lubchenco, statement to Presidential Commission co-chairs, October 7, 2010 (www.noaanews.noaa.gov/stories2010/PDFs/image2010-10-07-185425.pdf).

10. Personal interview with William Reilly, November 20, 2014.

11. Presidential Commission Report, p. 90.

12. Ibid., pp. 56–57.

13. U.S. Department of the Interior, Office of the Inspector General, Report of Investigation—Federal Moratorium on Deepwater Drilling, Case No. PI-PI-10-0562-1.

14. Ibid.

15. Ibid.

16. Ibid.

17. "Jody Freeman to Return in March after Serving in the White House," *Harvard Law Today*, February 24, 2010 (http://today.law.harvard.edu/jody-freeman-to-return-in-march-after-serving-in-the-white-house/?redirect=1).

18. In a 2014 study published by the National Bureau of Economic Research, a Harvard public policy professor found that there had been fewer than 400 claims for displaced rig support workers since very few rigs wound up leaving the Gulf despite the moratorium (the same study concluded that for a number of reasons the disaster as a whole resulted in little adverse impact on the labor market in Louisiana). See Joseph Aldy, "The Labor Market Impacts of the 2010 Deepwater Horizon Oil Spill and Offshore Oil Moratorium," NBER Working Paper 2040 (Cambridge, Mass.: National Bureau of Economic Research, August 2014), p. 24.

19. Ann Gerhart, "BP Chairman Talks about the 'Small People,' Further Angering the Gulf," *Washington Post*, June 16, 2010.

20. This section is based on an interview with a confidential congressional source.

21. Personal interview with Frances Beinecke, June 16, 2015.

22. Stephanie Condon, "Joseph Cao Tells BP Exec: In Samurai Days, You'd Kill Yourself," CBS News, June 15, 2010 (www.cbsnews.com/news/joseph-cao-tells-bp-exec-in-samurai-days-youd-kill-yourself/).

23. "The Role of BP in the Deepwater Horizon Explosion and Oil Spills," House of Representatives Subcommittee on Oversight and Investigations, Committee on Energy and Commerce, June 17, 2010 (http://tenc.net/a/hayward_hearing_trans.pdf).

24. Ibid.

25. Ibid.

26. Committee staffer, confidential report.

CHAPTER 5

1. Department of Justice, May 18, 2015, response to Freedom of Information Act Request No. 2015-01671 (by author).

2. Ibid.

3. Terry Carter, "Master of Disasters: Is Ken Feinberg Changing the Course of Mass Tort Resolution?," *ABA Journal*, January 2011.

4. Ken Feinberg, remarks made during visit to author's class, American University, November 8, 2011.

5. Manuel Roig-Franzia, "Kenneth Feinberg Talks about Compensation for Sept. 11, BP Oil Spill Victims," *Washington Post*, June 26, 2012 (www.washingtonpost.com/lifestyle/style/kenneth-feinberg-talks-about-compensation-for-sept-11-bp-oil-spill-victims/2012/06/26/gJQAZHLQ5V_story.html).

6. Report by the Claims Administrator of the Deepwater Horizon Economic and Property Damages Settlement Agreement on the Status of Claims Review, MDL No. 2179 (E.D. La), filed June 1, 2016.

7. "Morrell to Lead New BP America Communications and External Affairs Team," BP Global, September 5, 2013 (www.bp.com/en/global/corporate/press/press-releases/morrell-lead-new-bp-america-communications-external-affairs.html).

8. Campbell Robertson and Eric Lipton, "BP Is Criticized Over Oil Spill, but U.S. Missed Chances to Act," *New York Times*, April 30, 2010 (www.nytimes.com/2010/05/01/us/01gulf.html?pagewanted=all&_r=0).

9. "Morrell to Lead New BP America Communications and External Affairs Team," BP press release, September 5, 2013.

10. Alexandra Bruell, BP's Latest PR Push: Don't Take Advantage of Our Settlement Agreement," *Advertising Age,* August 23, 2013 (http://adage.com/article/media/bp-s-latest-pr-push-advantage/243815/).

11. *In re: Deepwater Horizon,* "Order & Reasons [Responding to Remand of Business Economic Loss Issues]," December 24, 2013 (www.laed.uscourts.gov/sites/default/files/OilSpill/Orders/12242013Order(RevisedBELremand).pdf).

12. *In re: Deepwater Horizon,* 739 F.3d 790, 795 (5th Cir.2014); 744 F.3d 370 (5th Cir. 2014), cert. denied 135 S.Ct. 754 (2014).

13. "Did '60 Minutes' Whitewash the Deepwater Horizon Oil Spill?," *Los Angeles Times,* May 7, 2014. The *60 Minutes* episode aired on May 4, 2014 and a transcript can be found here: www.cbsnews.com/news/over-a-barrel-bp-oil-spill-settlement-60-minutes/.

14. BDO Consulting, "Independent Evaluation of the Gulf Coast Claims Facility," April 19, 2012, Executive Summary, p. 2 (www.justice.gov/iso/opa/resources/697201241917226179477.pdf).

15. Independent External Investigation of the Deepwater Horizon Court Supervised Settlement Program Report of Special Master Louis J. Freeh, September 6, 2013, filed in *in re deepwater horizon.*

16. Order and Reasons, *In re Deepwater Horizon,* April 29, 2014.

17. *In re: Deepwater Horizon,* Order [Concerning the Special Master's Report of September 6, 2013; Imposing Certain Sanctions], February 26, 2015 (www.laed.uscourts.gov/sites/default/files/OilSpill/Orders/2262015Order(Sanctions).pdf).

18. Press Release, Department of Justice, U.S. Attorney's Office, Eastern District of Louisiana, October 28, 2015 (www.fbi.gov/contact-us/field-offices/neworleans/news/press-releases/slidell-man-sentenced-for-making-more-than-355-000-in-false-claims-to-the-deepwater-horizon-economic-claims-center).

19. *In re: Deepwater Horizon,* January 14, 2014, Report of Special Master Louis J. Freeh.

20. *In re: Deepwater Horizon,* 793 F.3d 479, 488 n. 8 (5th Cir., 2015), July 16, 2015.

21. Grand Jury Indictment, *United States v. Mikal C. Watts, et al.,* No. 1:15CR65LG. (S.D. Miss., September 15, 2015).

22. Ibid.

23. Ibid.

24. Associated Press, "The Latest: Oil Spill Claims Head Testifies in Fraud Trial," ABC News, July 20, 2016 (http://abcnews.go.com/US/wireStory/latest-trial-bp-oil-spill-fraud-case-40741222).

25. Anita Lee, "Attorney Watts, Law Firm Vindicated in Federal BP Fraud Case," *Sun Herald,* August 18, 2016.

26. Ibid.

27. *United States v. Montgomery* (No. 13-2596) (6th Cir. 2014) (unpublished opinion).

28. Ibid., p. 2.

29. Ibid., pp. 2–3.

30. Ibid., p. 15.

31. Press Release, Department of Justice, U.S. Attorney's Office, Northern District of Alabama, January 26, 2015, "Conspirators in Gulf Oil-Spill Fund Fraud Sentenced" (www.justice.gov/usao-ndal/pr/conspirators-gulf-oil-spill-fund-fraud-sentenced).

32. Press Release, Department of Justice, U.S. Attorney's Office, Northern District of Alabama, Wednesday, April 30, 2014, "Five Indicted for Conspiracy to Defraud Gulf Oil Spill Fund" (www.justice.gov/usao-ndal/pr/five-indicted-conspiracy-defraud-gulf-oil-spill-fund).

33. "Feds Hit Hard against Those Making False BP Claims, Maybe Too Hard," *Forbes*, July 15, 2014 (www.forbes.com/sites/walterpavlo/2014/07/15/feds-hit-hard-against-those-making-false-bp-claims-maybe-too-hard/#dbd3485aff30).

CHAPTER 6

1. Plea Agreement, United States v. BP Exploration and Production, Inc. (E.D. La.) (www.justice.gov/iso/opa/resources/43320121115143613990027.pdf). "Seaman's Manslaughter" is technically called "Misconduct or Neglect of Ship Officers." 18 U.S.C. § 1115. For a full discussion of other conceivable criminal charges, see David Uhlmann, "After the Spill Is Gone: The Gulf of Mexico, Environmental Crime, and the Criminal Law," *Michigan Law Review* 109 (2011).

2. Ibid.; see also *United States v. BP Exploration and Production, Inc.*, Reasons for Accepting Plea Agreement, No. 12-292, (E.D. La.) (January 29, 2013).

3. As of 2011, only 15 states had prosecuted corporations for manslaughter or criminally negligent homicide. See "Corporate Criminal Liability for Homicide: A Statutory Framework," *Duke Law Journal* 61, pp. 123, 133.

4. Corporate Manslaughter and Homicide Act of 2007.

5. See William J. Maakestad, "*State v. Ford Motor Co.*: Constitutional, Utilitarian and Moral Perspectives," *St. Louis University Law Journal* 27 (1983), p. 857 (explaining that the case was complicated by a change in Indiana law after the design and manufacture of the vehicle).

6. The president and managers of the company were convicted of murder, but their convictions were overturned on appeal on the grounds that the company had only been convicted of manslaughter. Although the murder conviction was reversed on appeal as being inconsistent with the manslaughter conviction of the company, the appellate court made clear that the evidence did not bar a retrial on the individual murder counts. *People v. O'Neil* 550 N.E.2d 1090 (Ill. App. 1990).

7. An excellent analysis of this debate appears in Duke law professor Sara Beale's "A Response to the Critics of Corporate Criminal Liability," *American Criminal Law Review* 46 (2009), p. 1491.

8. An example used in the classroom: If you're driving down the road within the speed limit and don't have ear buds in and you're not texting, and you hit somebody you didn't see, then you are liable for negligence. But it is gross negligence if you are speeding and your ear buds are in and all the warning lights are lighted up on your dashboard and you ignore them. See Steven Mufson, "BP, Halliburton, Transocean, Plaintiffs' Attorneys All Prepare to Face off in Gulf Oil Spill Trial," *Washington Post*, February 23, 2013 (www.washington post.com/business/economy/in-gulf-of-mexico-oil-spill-trial-bp-halliburton-transocean-plaintiffs-lawyers-prepare-to-face-off/2013/02/23/84aea08c-7d2a-11e2-82e8-61a46c2cde3d_story.html).

9. Personal interview with Jane Lubchenco, January 3, 2015.

10. Press Release, U.S. Department of Justice, February 14, 2013 (www.justice.gov/opa/pr/transocean-pleads-guilty-sentenced-pay-400-millionin-criminal-penalties-criminal).

11. Press Release, U.S. Department of Justice, September 19, 2013 (www.justice.gov/opa/pr/halliburton-pleads-guilty-destruction-evidence-connection-deepwater-horizon-disaster-and)

12. Press Release, U.S. Department of Justice, April 24, 2012, "Former BP Engineer Arrested for Obstruction of Justice in Connection with the Deepwater Horizon Criminal Investigation" (www.justice.gov/opa/pr/former-bp-engineer-arrested-obstruction-justice-connection-deepwater-horizon-criminal).

13. Press Release, U.S. Department of Justice, December 18, 2013, Former BP Engineer Convicted for Obstruction of Justice in Connection with the Deepwater Horizon Criminal Investigation (www.justice.gov/opa/pr/former-bp-engineer-convicted-obstruction-justice-connection-deepwater-horizon-criminal).

14. *United States v. Mix, Order and Reasons,* Criminal Action No. 12-171 (E.D. La. Feb. 13, 2014). The judge noted in a footnote that it was his understanding that BP would pay for Mix's defense provided he was acquitted.

15. *United States v. Mix,* 25 F. Supp.3d 914 (E.D. La. 2014).

16. *United States v. Mix,* 791 F.3d 603 (5th Cir. 2015).

17. Walter Pavlo, "Government Drops Obstruction Charges Against Former BP Engineer Kurt Mix," *Forbes,* Nov. 6, 2015.

18. Kurt Mix, "I Was an Oil Spill Scapegoat," *Wall Street Journal*, November 8, 2015.

19. Indictment for Obstruction of Congress and False Statements, United States v. David Rainey, November 14, 2012 (E.D. La) (www.justice.gov/iso/opa/resources/278201211151436583028449.pdf).

20. *United States v. Rainey,* 757 F.3d 234, 236, 248-49 (5th Cir. 2014).

21. Superseding Indictment for Involuntary Manslaughter, Seaman's Manslaughter, and Clean Water Act. *United States v. Robert Kaluza and Donald Vidrine.* Criminal No. 12-265. (E.D. La.) (www.justice.gov/iso/opa/resources/2520121115143638743323.pdf).

22. Indictment for Obstruction of Congress and False Statements, *United States v. David Rainey,* November 14, 2012 (E.D. La) (www.justice.gov/iso/opa/resources/2782012111514365832844.pdf).

23. Margaret Cronin Fisk and Laurel Brubaker Calkins, "BP Well-Site Managers' Oil-Spill Manslaughter Case Dropped," *Bloomberg,* December 2, 2015.

24. 176 F.3d 1116 (9th Cir. 1999), *cert. denied,* 528 U.S. 1102 (2000).

25. Press Release, U.S. Department of Justice, September 13, 2013 (www.justice.gov/opa/pr/halliburton-pleads-guilty-destruction-evidence-connection-deepwater-horizon-disaster-and).

26. Press Release 2012-231, SEC, November 15, 2012.

27. Complaint, *SEC v. Seilhan.* No. 2:14-cv-00893-CJF-SS. April 17, 2014 (www.sec.gov/litigation/complaints/2014/comp-pr2014-77.pdf).

28. Rena Steinzor and Anne Havemann, "Too Big to Obey: Why BP Should Be Debarred," *William and Mary Environmental Law and Policy Review* 36 (2011), pp. 81, 83.

29. Letter from Richard Pelletier, EPA, to Robert Dudley, BP, November 28, 2012.

30. Brief of the Government of the United Kingdom of Great Britain and Northern Ireland as Amicus Curiae in Support of Plaintiff's Motion for Summary Judgment, *BP Exploration and Production, et al. v. McCarthy,* No. 4:13-cv-2349 (S.D. Tex.), filed December 2, 2013.

31. *Hornbeck Offshore Services v. Salazar,* 696 F.Supp.2d 627 (E.D. La. 2010).

32. Ibid., p. 639. In a July 2010 study funded by the American Energy Alliance, a Louisiana State University Business School professor estimated that the total cost of the moratorium to the region to be $2.1 billion, including $500 million in lost wages. See Joseph R. Mason, "The Economic Cost of a Moratorium on Offshore Oil and Gas Exploration to the Gulf Region," Louisiana State University, July 2010 (http://instituteforenergyresearch.org/wp-content/uploads/2010/07/Mason-Economic_Cost_of_Offshore_Moratorium.pdf).

33. *Hornbeck Offshore Services v. Salazar,* p. 631.

34. Joel Cohen, *Blindfolds Off: Judges on How They Decide* (American Bar Association, 2014), p. 59.

35. *Cobell v. Babbitt,* 37 F. Supp. 2d 6 (D.D.C. 1999); *Cobell v. Norton,* 226 F. Supp. 2d 1 (D.D.C. 2002), vacated in part, *Cobell v. Norton,* 334 F.3d 1128 (D.C. Cir. 2003). See Daniel Jacobs, "The Role of the Federal Government in Defending Public Interest Litigation," *Santa Clara Law Review* 44 (2003), p. 1.

36. *Hornbeck v. Salazar*, 713 F.3d 787 (5th Cir. 2013), *cert. denied sub nom Hornbeck v. Jewell*, 134 S.Ct. 823 (2013).

37. See "The Judge behind the Ruling," *Wall Street Journal*, June 24, 2010.

38. *Ensco Offshore Co. v. Salazar* (No. 10-1941, ED LA).

39. *Ensco Offshore Co. v. Salazar* (No. 11-30491, 5th Cir. 2011).

40. See www.judicialwatch.orgwp-content/uploads/2013/11/C-L-Feldman-Martin-Financial-Disclosure-Report-for-2010.pdf. See also Charlie Savage, "Drilling Ban Blocked; U.S. Will Issue New Order," *New York Times,* June 22, 2010.

41. Ibid. (emphasis added).

42. Both the official docket of the court and the hearing transcript itself reflect that the hearing took place on June 21, 2010. Reviewed on PACER (the U.S. court document system), November 21, 2014.

43. See www.judicialwatch.orgwp-content/uploads/2013/11/C-L-Feldman-Martin-Financial-Disclosure-Report-for-2010.pdf.

44. "Allis-Chalmers Energy Announces Early BOP Recertification Plan," June 30, 2010 (www.businesswire.com/news/home/20100630006596/en/Allis-Chalmers-Energy-Announces-Early-BOP-Recertification-Plan#.VFFPemddUuc).

45. Ibid.

46. See Code of Conduct for United States Judges, Canon 2A Commentary, "A judge must avoid all impropriety and appearance of impropriety."

47. Cohen, *Blindfolds Off*, p. 65.

CHAPTER 7

1. United States Judicial Panel on Multidistrict Litigation, Transfer Order, No. MDL 2179, August 10, 2010.

2. *In re: Oil Spill by the Oil Rig Deepwater Horizon in the Gulf of Mexico on April 20, 2010*, United States District Court, Eastern District of Louisiana, Transcript of Nonjury Trial Proceedings Heard before the Honorable Carl J. Barbier, United States District Judge, February 25, 2012 (Day One, morning), p. 27 (www.mdl2179trialdocs.com/releases/release201302250700001/2013-02-25_BP_Trial_Day_01_AM-Final.pdf).

3. Ibid., pp. 27–28.

4. Ibid., p. 65.

5. Ibid., p. 70.

6. Ibid., p. 64.

7. Ibid., p. 66.

8. Ibid., p. 77.

9. Ibid., pp. 78–79.

10. Ibid., p. 79.

11. Ibid., p. 80.

12. Ibid., p. 81.

13. Ibid., p. 78.

14. Ibid., p. 151.

15. Ibid., p. 240.

16. Trial transcript, Phase One, Day Two, p. 482 (www.mdl2179trialdocs.com/index.php?page=details&release_id=201302261000002).

17. Trial transcript, Phase One, Day Two, p. 375.

18. Phase One Decision.

19. Trial transcript, Phase One, Day Three, p. 671 (www.mdl2179trialdocs.com/releases/release201302271030003/2013-2-27_BP_Trial_Day_03_AM-Final.pdf).

20. Phase One Decision, 21 F. Supp. 3d, p. 674.

21. Trial transcript, Phase One, Day 23, pp. 7443, 7444, 7446 (www.mdl2179trialdocs.com/releases/release201304080900023/2013-04-08_BP_Trial_Day_23_AM-Final.pdf).

22. Phase One Decision, p. 675.

23. Trial transcript, Phase One, Day Seven, p. 2086 (www.mdl2179trialdocs.com/releases/release201303060945007/2013-03-06_BP_Trial_Day_7_AM-Final.pdf).

24. Trial transcript, Phase One, Day Two, p. 358.

25. Trial transcript, Phase One, Day 26, p. 8648. See also "Louisiana AG Political Contributors Reap Lucrative Legal Contracts in BP Litigation; Biggest Contractor, Caught Napping in Court, Has Reaped $12M," *Louisiana Record*, April 23, 2015.

26. Post-Trial Brief of Defendants BP Exploration & Production, Inc., et al. (filed June 21, 2013), p. 3.

27. Ibid., p. 2.

28. United States of America's Post-Trial Brief for Phase One (filed June 21, 2013), p. 12.

29. Ibid., p. 22.

30. Phase One Decision, p. 657.

31. Ibid., pp. 737–43.

32. Ibid., pp. 741-43.

33. Ibid., p. 742.

34. Ibid., p. 757.

35. Findings of Fact and Conclusions of Law, Phase Two Trial, January 15, 2015, p. 40 (www.laed.uscourts.gov/sites/default/files/OilSpill/Orders/1152015FindingsPhaseTwo.pdf).

36. Ibid., pp. 43–44.

37. 33 U.S.C. sec. 1321(b)(8).

38. Trial transcript, Phase Three, Day One, p. 63 (www.mdl2179trialdocs.com/releases/release201501201000001/2015-01-20_AM_Opening_and_Adm_Austin.pdf).

39. Trial transcript, Phase Three, Day Two, p. 336 (www.mdl2179trialdocs.com/releases/release201501211000002/2015-01-21_AM_Boesch.pdf).

40. Trial transcript, Phase Three, Day Three, p. 977 (www.mdl2179trialdocs.com/releases/release201501220800003/2015-01-22_PM_Walkup_Suttles_Video_Bray_Video_and_Quivik.pdf).

41. Department of Justice, "Fact Sheet on Agreement in Principle with BP," July 2, 2015 (www.justice.gov/opa/file/625141/download).

42. Department of Justice, "Attorney General Loretta E. Lynch Delivers Remarks at Press Conference Announcing Settlement with BP to Resolve Civil Claims Over Deepwater Horizon Oil Spill," October 5, 2015 (www.justice.gov/opa/speech/attorney-general-loretta-e-lynch-delivers-remarks-press-conference-announcing-settlement).

43. Consent Decree Among Defendant BP Exploration & Production Inc. ("BPXP"), the United States of America, and the States of Alabama, Florida, Louisiana, Mississippi, and Texas, April 4, 2016 (www.laed.uscourts.gov/sites/default/files/OilSpill/4042016ConsentDecree_0.pdf); and Final Judgment (www.laed.uscourts.gov/sites/default/files/OilSpill/4042016Judgment%28reConsentDecree%29_0.pdf).

CHAPTER 8

1. The $17 billion that could be absorbed by taxpayers is based on the deductible portion of BP's *total* expenses (at a roughly 33 percent rate) of $61.6 billion as of July 2016.

2. "BP Estimates All Remaining Material Deepwater Horizon Liabilities," BP, July 14, 2016 (www.bp.com/en/global/corporate/press/press-releases/bp-estimates-all-remaining-material-deepwater-horizon-liabilitie.html).

3. Denis P. Boudreaux, Spuma Rao, Praveen Das, and Nancy Rumore, "How Much Did the Gulf Spill Actually Cost British Petroleum Shareholders?," *Journal of International Energy Policy*, Spring 2013 (www.cluteinstitute.com/ojs/index.php/JIEP/article/view/7891/7950).

4. Christine Hauser, "BP Shareholders Take It on the Chin," *New York Times*, June 16, 2010 (www.nytimes.com/2010/06/17/business/energy-environment/17investors.html?_r=0).

5. *Ludlow v. BP*, 800 F.3d 674 (5th Cir. 2015).

6. "Deepwater Horizon—MDL 2185 Securities Litigation," BP press release, June 3, 2016 (www.bp.com/en/global/corporate/press/press-releases/deepwater-horizon-mdl-2185-securities-litigation.html).

7. Seahawk Drilling and ATP are the two companies that reportedly attributed their bankruptcies to the moratorium. See Margaret Cronin Fisk and Laurel Brubaker Calkins, "BP Avoids Lawsuits over Moratorium That Followed Gulf Spill," *Bloomberg,* March 10, 2016 (www.bloomberg.com/news/articles/2016-03-10/bp-won-t-face-moratorium-claims-over-oil-spill-judge-says-ilmwrcpz), as well as Ryan Lizza, "As the World Burns," *The New Yorker,* October 11, 2010.

8. United States District Court for the Eastern District of Louisiana, "Memorandum in Support of Unopposed Motion by the United States for Entry of Consent Decree with BP," March 22, 2016, p. 28 (www.justice.gov/enrd/file/834471/download) (hereafter, "Memorandum in Support of Entry of Consent Decree").

9. In a note BP disclosed, "The interest rate will be fixed at the average market yield on U.S. Treasury securities at 2-year and 3-year constant maturities, quoted on an investment basis by the US Federal Reserve (H. 15 Release), for the period from 28 May 2014 to 27 May 2015." See "BP to Settle Federal, State and Local Deepwater Horizon Claims for up to $18.7 Billion with Payments to Be Spread over 18 Years," BP press release, July 2, 2015 (www.bp.com/en/global/corporate/press/press-releases/bp-to-settle-federal-state-local-deepwater-horizon-claims.html).

10. EPA, "Guidance on Evaluating a Violator's Ability to Pay a Civil Penalty in an Administrative Enforcement Action," June 29, 2015, pp. 17–18 (www.epa.gov/sites/production/files/2015-06/documents/atp-penalty-evaluate-2015.pdf).

11. EPA, Interim Clean Water Act Settlement Penalty Policy, March 1, 1995, p. 21.

12. Michael Conathan, "The True Value of BP's $18.7-Billion Settlement," Center for American Progress, July 30, 2015 (www.americanprogress.org/issues/green/report/2015/07/30/118316/the-true-value-of-bps-18-7-billion-settlement/). This report calculated the applied difference between the agreed-upon interest rate and market rates at $2 billion. However, it apparently based that figure on BP's total valuation of the settlement ($18.7 billion), whereas interest payments were due only on the $5.5 billion civil penalty and $7.1 billion natural resource damage payments.

13. Memorandum in Support of Entry of Consent Decree, Department of Justice, p. 21, n. 28.

14. Ibid., p. 21.

15. The city of Birmingham, Alabama, decided to spend $200,000 of the roughly $1 million it received in the BP settlement to promote football. The money went to support the Birmingham Bowl game that took place on December 30, 2015. The Auburn Tigers beat the Memphis Tigers 31–10.

16. David P. Batker, Isabel de la Torre, Robert Costanza, Paula Swedeen, John W. Day Jr., Roelof Boumans, and Kenneth Bagstad, *Gaining Ground—Wetlands, Hurricanes and the Economy: The Value of Restoring the Mississippi River Delta* (Tacoma, Wash.: Earth Economics, 2010).

17. Robert Costanza, David Batker, John W. Day, Jr., Rusty A. Feagin, M. Luisa Martinez, and Joe Roman. "The Perfect Spill: Solutions for Averting the Next *Deepwater Horizon*." *Solutions,* 1: 5 (2010).

CHAPTER 9

1. Interview with confidential source.

2. Presidential Commission Report, p. 84 (www.gpo.gov/fdsys/pkg/GPO-OILCOMMISSION/pdf/GPO-OILCOMMISSION.pdf).

3. Remarks by the President on Energy Security at Andrews Air Force Base, 3/31/2010, White House (www.whitehouse.gov/the-press-office/remarks-president-energy-security-andrews-air-force-base-3312010).

4. Ryan Lizza, "As the World Burns," *The New Yorker*, October 11, 2010.

5. Interview with Carol Browner, "The Spill," PBS Frontline (www.pbs.org/wgbh/pages/frontline/the-spill/interviews/carol-browner.html#1).

6. Cabinet Meeting on BP Oil Spill, White House Press Room, June 7, 2010 (www.whitehouse.gov/photos-and-video/video/cabinet-meeting-bp-oil-spill).

7. Margaret Cronin Fisk and Laurel Brubaker Calkins, "BP Avoids Lawsuits over Moratorium That Followed Gulf Spill," *Bloomberg,* March 10, 2016 (www.bloomberg.com/news/articles/2016-03-10/bp-won-t-face-moratorium-claims-over-oil-spill-judge-says-ilmwrcpz).

8. In 1991 President George H. W. Bush announced Operation Desert Storm—the start of the first Gulf War—from the Oval Office. Some ten years later, his son, President George W. Bush, would give his first Oval Office address to the nation on 9/11. Having first learned of the attacks while reading to children at a Florida elementary school, Bush flew first to a Louisiana Air Force base and then to a Nebraska nuclear bunker before returning to Washington that evening only shortly before the speech.

9. Remarks by the President to the Nation on the BP Oil Spill, White House, June 15, 2010 (www.whitehouse.gov/the-press-office/remarks-president-nation-bp-oil-spill).

10. Ibid.

11. Ann Finkbeiner, *The Jasons: The Secret History of Science's Postwar Elite* (New York: Viking, 2007). See also Berkeley SESPA (Scientists and Engineers for Social and Political Action), "Science Against the People: The Story

of Jason . . . ," December 1972 (http://socrates.berkeley.edu/~schwrtz/SftP/
JASON/Jason.html).

12. David Biello, "How Science Stopped BP's Gulf of Mexico Oil Spill,"
Scientific American, April 19, 2011.

13. Katie Howell, "As Bill Clinton Joins 'Bomb the Well' Club, Experts Wince,"
New York Times, June 30, 2010 (www.nytimes.com/gwire/2010/06/30/30green
wire-as-bill-clinton-joins-bomb-the-well-club-exper-50220.html).

14. The council's website is at https://restorethegulf.gov.

15. Remarks by the President to the Nation on the BP Oil Spill.

16. "From the Oval Office," *New York Times* editorial, June 15, 2010 (www.
nytimes.com/2010/06/16/opinion/16wed1.html?_r=0).

17. Remarks by the President to the Nation on the BP Oil Spill.

18. See, for example, "Gulf Oil Spill: White House Says BP Containment a Turning Point," ABC News, *Good Morning America*, August 4, 2010 (http://abcnews.
go.com/GMA/video/white-house-turning-point-oil-containment-11320458).

19. E-mail from Jane Lubchenco, August 4, 2010, 8:45 A.M.

20. Personal interview with Jane Lubchenco, January 3, 2015.

21. See Press Briefing by Press Secretary Robert Gibbs, Admiral Thad Allen,
Carol Browner, and Dr. Lubchenco, White House, August 4, 2010 (www.white
house.gov/the-press-office/press-briefing-press-secretary-robert-gibbs-admiral-
thad-allen-carol-browner-and-dr).

22. Ibid.

23. Presidential Commission Report, pp. 167–69.

24. Presidential Commission Staff Working Paper No. 3, originally
released 10/6/10; updated 1/11/11 (http://cybercemetery.unt.edu/archive/oil-
spill/20121211011123/http://www.oilspillcommission.gov/sites/default/files/
documents/Updated%20Amount%20and%20Fate%20of%20the%20Oil%20
Working%20Paper.pdf).

25. Ibid.

26. Personal interview with Jane Lubchenco, January 3, 2015.

27. "Poll Shows Negative Ratings for BP, Federal Government," *Washington Post,* "Behind the Numbers," undated (http://voices.washingtonpost.com/
behind-the-numbers/2010/06/poll_shows_negative_ratings_fo.html).

28. Karl Rove, "Yes, the Gulf Spill Is Obama's Katrina," *Wall Street Journal*,
May 27, 2010 (www.wsj.com/articles/SB1000142405274870471700457526877
52362770856).

29. "President Arrives in Alabama, Briefed on Hurricane Katrina," White
House, September 2, 2005 (http://georgewbush-whitehouse.archives.gov/news/
releases/2005/09/images/20050902-2_f1g5125-515h.html).

30. Adam Nagourney, "Lost Horizons," *New York Times Magazine*, September 24, 2006 (www.nytimes.com/2006/09/24/magazine/24melman.html?_r=0).

31. Carol Browner and John Podesta, "Why We Now Oppose Drilling in the Arctic," *Bloomberg View*, January 17, 2013 (www.bloomberg.com/news/2013-01-17/why-we-now-oppose-drilling-in-the-arctic.html).

32. Personal interview, confidential source, November 6, 2014.

33. "BP Oil Spill: I'd Like My Life Back," *The Guardian*, June 1, 2010.

34. *United States v. BP Exploration and Production,* Information for Seaman's Manslaughter, Clean Water Act, Migratory Bird Treaty Act, and Obstruction of Congress (E.D. La., November 15, 2012) (www.justice.gov/iso/opa/resources/73920121115143627533671.pdf).

35. Ibid.

36. Ibid. Also see Ben Casselman and Keith Johnson, "Coast Guard Log Details Early Hours of Spill," *Wall Street Journal*, June 4, 2010 (www.wsj.com/articles/SB10001424052748703340904575285381540974958).

37. Press Release, Department of Justice, November 15, 2012, "BP Exploration and Production Inc. Agrees to Plead Guilty to Felony Manslaughter, Environmental Crimes and Obstruction of Congress Surrounding Deepwater Horizon Incident" (www.justice.gov/opa/pr/bp-exploration-and-production-inc-agrees-plead-guilty-felony-manslaughter-environmental).

38. Yet another BP commercial referring to "stop job" authority ran during the Republican National Convention in July 2016: "BP Safety: Stop the Job" (www.youtube.com/watch?v=QGZNnAQHldw).

CHAPTER 10

1. "Reasons for Accepting Plea Agreement," *United States v. BP Exploration and Production, Inc.*, No. 12-292 (E.D. La., January 29, 2013), p. 5. "BP" is used here to refer to any BP company in the "BP family," past or present.

2. Press Release, Department of Justice, February 7, 1995.

3. "An Environmental Management System is a set of processes and practices that enable an organization to reduce its environmental impacts and increase its operating efficiency," at Environmental Management Systems (EMS), Environmental Protection Agency (www.epa.gov/ems).

4. Press Release, Department of Justice, September 23, 1999.

5. Press Release, Environmental Protection Agency, January 19, 2001 (https://yosemite.epa.gov/opa/admpress.nsf/b1ab9f485b098972852562e7004d-c686/1837d58f6a4da14e852569d900633bef?OpenDocument).

6. Report of the BP U.S. Refineries Independent Safety Review Panel, January 2007 (http://sunnyday.mit.edu/Baker-panel-report.pdf).

7. Press Release, Department of Justice, February 19, 2009 (www.justice.gov/opa/pr/bp-products-pay-nearly-180-million-settle-clean-air-violations-texas-city-refinery).

8. Press Release, Department of Justice, October 25, 2007.

9. "BP North Slope Spill Reveals A History of Substandard Environmental Performance: A Preliminary Report to the Alaska Forum for Environmental Responsibility," Richard A. Fineberg, March 15, 2006 (www.finebergresearch. com/pdf/Neport060315Rev.pdf).

10. Wesley Loy, "BP settles over Prudhoe Bay Spills," *Alaska Dispatch News*, August 2, 2014.

11. "Climate Change Speech by John Browne, Group Chief Executive, British Petroleum (BP)," Stanford University, May 19, 1997 (http://dieoff.org/page106. htm).

12. Intergovernmental Panel on Climate Change, "Methodological and Technology Issues in Technology Transfer: A Special Report of IPCC Working Group III" (Cambridge University Press, 2000), p. 205.

13. "Climate Change Speech by John Browne."

14. David G. Victor and Joshua C. House, "BP's Emissions Trading System," *Energy Policy* 34 (2006), pp. 2100–12.

15. Ibid.

16. Ibid.

17. "2002 World Summit Greenwash Academy Awards Program," August 23, 2002 (www.foe.co.uk/sites/default/files/downloads/summit_greenwash_ awards.pdf). In its 1998 merger with Amoco, BP acquired a 50 percent stake in a solar company, Solarex, and it bought out Enron's 50 percent share the following year. Solar represented a very small piece of BP's business holdings.

18. Richard Wray, "BP Fails to Reduce Greenhouse Gases," *The Guardian*, April 11, 2005 (www.theguardian.com/business/2005/apr/12/oilandpetrol. climatechange).

19. Shanta Barley, "BP Brings Green Era to a Close," BBC News, May 11, 2009 (http://news.bbc.co.uk/2/hi/science/nature/8040468.stm).

20. "BP Wins Coveted 'Emerald Paintbrush' Award for Worst Greenwash of 2009," Greenpeace, December 22, 2009 (www.greenpeace.org/international/ en/news/Blogs/makingwaves/bp-wins-coveted-emerald-paintbrush-award-for-/ blog/10204/).

21. "The Spill," *PBS Frontline*, October 26, 2010.

22. Lynn Stout, *The Shareholder Value Myth* (San Francisco: Berrett-Koehler Publishers, 2012), pp. 3–4.

23. Ibid., p. 87.

24. Personal interview with Frances Beinecke, June 16, 2015.

25. Dan Esty and Andrew Winston, *Green to Gold: How Smart Companies Use Environmental Strategy to Innovate, Create Value, and Build Competitive Advantage* (Yale University Press, 2006).

26. Stan Cox, "Capitalism's New Mantra: Eat Your Greens," *Business Day*, May 3, 2008.

27. Dan Esty and Andrew Winston, *Green to Gold*, paperback ed. (Hoboken, N.J.: John Wiley & Sons, 2009), p. xii.

28. *Green to Gold* (2009), p. 25. Despite the disclaimer that the rankings today would be different, the paperback edition still contains the original list with BP ranked as the number one international "WaveRider."

29. Ibid., p. 141.

30. Ibid., pp. 25–26.

31. See www.greenbiz.com/blog/2014/08/13/greenbiz-twitterati-2014.

CHAPTER 11

1. Personal interview with Jane Lubchenco, January 3, 2015.

2. Hannah Waters, "Breaking Down the Myths and Misconceptions About the Gulf Oil Spill," *Smithsonian Magazine*, April 17, 2014 (www.smithsonian mag.com/science-nature/clarifying-myths-and-misconceptions-about-gulf-oil-spill-180951136/?no-ist).

3. Testimony of Jeff Frandahl, executive director, National Fish and Wildlife Foundation, before the Senate Committee on Commerce Regarding "Gulf Restoration: A Progress Report Three Years after the Deepwater Horizon Disaster," June 6, 2013.

4. Personal interview with Roger Helm, March 10–11, 2015.

5. Personal interview with David Valentine, February 17, 2015.

6. Personal interview with Bob Bendick, December 3, 2014.

7. "A Comprehensive Restoration Plan for the Gulf of Mexico," Deepwater Horizon Natural Resource Damage Assessment Trustees, February 19, 2016 (www.gulfspillrestoration.noaa.gov/restoration-planning/gulf-plan).

8. Ibid., section 1.5.2.

9. Ibid., section 1.5.3.

10. Juliet Eilperin, "BP Settlement a Boon to Conservation Group," *Washington Post*, November 12, 2012 (www.washingtonpost.com/national/health-science/bp-settlement-a-boon-to-conservation-group/2012/11/16/ddcb2790-302b-11e2-a30e-5ca76eeec857_story.html).

11. Ibid.

CHAPTER 12

The chapter title was inspired by George Ball's *New York Times Magazine* cover story "The Lessons of Vietnam: Have We Learned or Only Failed?" April 1, 1973 (www.nytimes.com/1973/04/01/archives/have-we-learned-or-only-failed-the-lessons-of-vietnam-vietnam.html).

1. Presidential Commission Report, p. 84.

2. Ibid., p. 225.

3. Press Release, Department of Justice, November 19, 2015 (www.justice.gov/opa/pr/three-companies-and-three-individuals-charged-fatal-2012-gulf-mexico-oil-drilling-platform).

4. "Top 10 Worst Cabinet Members," *Time* (http://content.time.com/time/specials/packages/completelist/0,29569,1858691,00.html).

5. Steven R. Weisman, "Watt Quits Post; President Accepts with 'Reluctance,'" *New York Times*, October 10, 1983 (www.nytimes.com/1983/10/10/us/watt-quits-post-president-accepts-with-reluctance.html).

6. David Johnston, "Ex-Interior Chief is Indicted in Influence-Peddling Case," *New York Times*, February 23, 1995 (www.nytimes.com/1995/02/23/us/ex-interior-chief-is-indicted-in-influence-peddling-case.html).

7. David Johnston, "Former Interior Secretary Avoids Trial with a Guilty Plea," *New York Times*, January 3, 1996 (www.nytimes.com/1996/01/03/us/former-interior-secretary-avoids-trial-with-a-guilty-plea.html).

8. Peter Jan Honigsberg, "Conflict of Interest That Led to the Gulf Oil Disaster," 41 *ELI Reporter* 10414 (2011).

9. Presidential Commission Report, p. 250.

10. Ibid., p. 225.

11. Ibid.

12. Personal interview with Frances Beinecke, June 16, 2015.

13. U.S. Department of the Interior, Office of the Inspector General, Report of Investigation, Island Operating Company, et al., Case No. PI-GA-09-0102-I, March 31, 2010, and cover memo April 12, 2010 (www.doioig.gov/sites/doioig.gov/files/IslandOperatingCo.pdf).

14. Ibid.

15. Oil Spill Commission Action, "Assessing Progress: Implementing the Recommendations of the National Oil Spill Commission," April 17, 2012 (http://oscaction.org/wp-content/uploads/OSCA-Assessment-report.pdf); Oil Spill Commission Action, "Assessing Progress: Three Years Later," 2013 (http://oscaction.org/osca-assessment-report-2013/).

16. Oil Spill Commission Action, "Assessing Progress," 2012, p. 2.

17. Bureau of Safety and Environmental Enforcement and Bureau of Ocean Energy Management, "Reforms since the Deepwater Tragedy" (www.boem.gov/Reforms-since-the-Deepwater-Horizon-Tragedy/).

18. Bureau of Ocean Energy Management, "Interior Issues Final Regulations to Raise Safety & Environmental Standards for Any Future Exploratory Drilling in U.S. Arctic Waters," July 7, 2016 (www.boem.gov/press07072016/).

19. Government Accountability Office, "Interior's Reorganization Complete, but Challenges Remain in Implementing New Requirements," Report 12-423, July 30, 2012 (http://gao.gov/products/GAO-12-423).

20. Jennifer A. Dlouhy, "Q&A: Top Regulator Sees Offshore Safety as a Work in Progress," *Houston Chronicle,* May 6, 2015 (www.houstonchronicle.com/

business/energy/conferences/article/Top-regulator-sees-offshore-safety-as-work-in-6247567.php).

21. Statement of Abigail Ross Hopper, director, Bureau of Ocean Energy Management, before the House Committee on Natural Resources, Subcommittee on Energy and Mineral Resources, March 17, 2015.

22. U.S. Chemical Safety Board, "The U.S. Chemical Safety Board's Investigation into the Macondo Disaster Finds Offshore Risk Management and Regulatory Oversight Still Inadequate in Gulf of Mexico," April 13, 2016 (www.csb.gov/the-us-chemical-safety-boards-investigation-into-the-macondo-disaster-finds-offshore-risk-management-and-regulatory-oversight-still-inadequate-in-gulf-of-mexico/).

23. Presidential Commission Report, p. 67.

24. Ibid.

25. Peter Lehner with Bob Deans, *In Deep Water: The Anatomy of a Disaster, the Fate of the Gulf, and How to End Our Oil Addiction* (New York: Or Books, 2010).

26. "Don't Blame BP—Gulf of Mexico Oil Spill Is America's Fault," *The Telegraph*, May 28, 2010 (www.telegraph.co.uk/finance/comment/jeremy-warner/7779914/Dont-blame-BP-Gulf-of-Mexico-oil-spill-is-Americas-fault.html).

27. See www.pewresearch.org/daily-number/offshore-drilling-support-one-year-after-gulf-oil-spill/ and www.pewinternet.org/2015/07/01/americans-politics-and-science-issues/pi_2015-07-01_science-and-politics_2-16/.

28. Personal interview with Bob Graham, May 3, 2015.

29. "Exploration and Production," BP, undated (www.bp.com/en_us/bp-us/what-we-do/exploration-and-production.html).

30. See Andrew Hopkins, *Disastrous Decisions: The Human and Organizational Causes of the Gulf of Mexico Blowout* (Sydney: CCH Australia, 2012).

INDEX

WITHDRAWN

PORTLAND PUBLIC LIBRARY
5 MONUMENT SQUARE
PORTLAND ME 04101

MANEW   07/25/2016